LANDING NATIVE FISHERIES

Law and Society Series
W. Wesley Pue, General Editor

The Law and Society Series explores law as a socially embedded phenomenon. It is premised on the understanding that the conventional division of law from society creates false dichotomies in thinking, scholarship, educational practice, and social life. Books in the series treat law and society as mutually constitutive and seek to bridge scholarship emerging from interdisciplinary engagement of law with disciplines such as politics, social theory, history, political economy, and gender studies.

A list of the titles in this series appears at the end of this book.

LANDING NATIVE FISHERIES

Indian Reserves and Fishing Rights in British Columbia,
1849-1925

DOUGLAS C. HARRIS

UBCPress · Vancouver · Toronto

16 15 14 13 12 11 10 09 08   5 4 3 2 1

Printed in Canada with vegetable-based inks on FSC-certified ancient-forest-free
paper (100% post-consumer recycled) that is processed chlorine- and acid-free.

Library and Archives Canada Cataloguing in Publication

Harris, Douglas C. (Douglas Colebrook)
   Landing native fisheries : Indian reserves and fishing rights in British Columbia,
1849-1925 / Douglas C. Harris.

(Law and society 1496-4953)
Includes bibliographical references and index.
ISBN 978-0-7748-1419-5

   1. Indians of North America – Fishing – Law and legislation – British Columbia –
History. 2. Fishery law and legislation – British Columbia – History. 3. Indians of
North America – Legal status, laws, etc. – British Columbia – History. 4. Indians
of North America – Land tenure – British Columbia – History. 5. Indians of North
America – Canada – Government relations. I. Title. II. Series: Law and society series
(Vancouver, B.C.)

KEB529.5.H8H373 2008   346.71104'6956   C2008-900087-0

Canadä

UBC Press gratefully acknowledges the financial support for our publishing program
of the Government of Canada through the Book Publishing Industry Development
Program (BPIDP), and of the Canada Council for the Arts, and the British Columbia
Arts Council.

This book has been published with the help of a grant from the Canadian Federation
for the Humanities and Social Sciences, through the Aid to Scholarly Publications
Programme, using funds provided by the Social Sciences and Humanities Research
Council of Canada, and with the help of the K.D. Srivastava Fund.

UBC Press
The University of British Columbia
2029 West Mall
Vancouver, BC V6T 1Z2
604-822-5959 / Fax: 604-822-6083
www.ubcpress.ca

Contents

Illustrations

Photographs

Tables

Acknowledgments

In 1996, the Supreme Court of Canada issued judgments in seven Aboriginal fishing rights cases, five of which originated in British Columbia and two that came from Quebec.[1] Following the court's 1990 landmark decision in *R. v. Sparrow*,[2] a case involving the fishing rights of the Musqueam that established the framework for the interpretation of Aboriginal rights in the Canadian Constitution, these seven cases confirmed the pivotal place of fishing rights in the general development of Aboriginal rights in Canada.

Most of my scholarship, which also began in earnest in 1996, has revolved around the five fishing rights cases from British Columbia. They were too intriguing in their historical interpretation, and too important in their legal developments, for a student interested in law, colonialism, and British Columbia to ignore. These court decisions drew me to the fisheries.

Although not much in evidence until the final few pages of my first book, the rulings of the Supreme Court provided much of the impetus for it. To understand both the emerging case law, so clearly indicative of continuing conflict between the state and First Nations over fisheries, and the role of law in that conflict, I undertook an exploration of the historical and legal context from which the litigation emerged. In *Fish, Law, and Colonialism*, I sought to describe that context in the late nineteenth century and, through several case studies, into the early twentieth.

That project, however, left me more than a few decades short of the cases I had set out to understand. In its writing, I also discovered there were other stories to be told or retold, one of which was an account of the connections between Indian reserves and Native fisheries, the subject of two of the Supreme Court decisions from 1996: *R. v. Nikal* and *R. v. Lewis*. These cases, and the historical record that they opened for me, led me to believe that Canada's regulation of the fisheries in British Columbia could be more fully understood in the context of the process of Indian reserve allotments. Similarly, the emergence of an Indian reserve geography could be better explained in the light of the fisheries and their regulation. Both these processes – the allotment of reserves and the regulation of the fisheries – occurred together in British Columbia and in tandem with the rise of the industrial/commercial fishery. It is this configuration – the emergence of the industrial/commercial fishery, the allotment of Indian reserves, and the introduction of Canadian fisheries law – that is the focus of this book.

The result is a book that, in terms of the fishing rights cases from 1996, speaks most directly to the issues in *Nikal* and *Lewis*. The documents that the lawyers assembled and the arguments they crafted in putting those cases before the Supreme Court of Canada were immensely helpful to me as I worked to explain the connections between Indian reserves and the fisheries. But although written in the shadow of litigation, decided and continuing, this book is neither a direct product of that litigation nor written specifically in response to it. Rather, the book is intended to address the need for a clearer understanding of the deep historical and legal currents that inform the continuing conflict over the fisheries on Canada's west coast. More generally, the book is also intended as a reflection on the role of law in the late-nineteenth- and early-twentieth-century settler colonialism of British Columbia.

A great many people have helped me over the past years as I worked on this project, and it is my considerable pleasure to thank them. The project began during my time at York University's Osgoode Hall Law School, a faculty blessed with an extraordinary collection of people. Doug Hay, whose work drew me to Osgoode, provided wonderful guidance and inspiration as my supervisor. Brian Slattery, Kent McNeil, and Gordon Christie were part of a larger community that inspired collegiality and critical inquiry. Eric Tucker as associate dean of Graduate Studies, Lea Dooley as graduate program coordinator, and Dean Peter Hogg, through their commitment to graduate studies and the graduate students, brought new life to the graduate program. Shin Imai, Shelley Gavigan, and Bill Wicken all provided critical and constructive feedback. My classmates, particularly Israel Doron and Mundy McLaughlin, were also instrumental in shaping a collegial environment of shared intellectual inquiry.

Midway through the project I accepted an assistant professor's position in the Faculty of Law at the University of British Columbia, and another group of remarkable scholars, some of whom had been my teachers, became my colleagues. Of them, Wes Pue and Ruth Buchanan deserve my particular thanks for many discussions on shared interests in law and geography, and June McCue and Michael Jackson for discussions of Aboriginal law. Michael Thoms, for much of this time a PhD student in history at UBC, engaged me in many discussions over the fisheries and provided extensive and insightful critical feedback on earlier drafts of this manuscript. Graeme Wynn of UBC's Geography Department gave me an opportunity to present portions of the manuscript in the Nature/History/Society lecture series. Dianne Newell, as a historian and as director of the Peter Wall Institute at UBC, broadened my interdisciplinary horizons through the Early Career Scholars program. Across the Strait of Georgia at the University of Victoria, Hamar Foster and John McLaren have been immensely helpful and interested, providing encouragement and opportunities to present my work. Rosemary Ommer, who directed the Coasts Under Stress project from the University of Victoria, found space for this study within that large, interdisciplinary research project. Chris Franks and the Law and Society Research Cluster at the University of Manitoba gave me an opportunity to discuss my work with them. Dan Marshall pointed me towards some important documents in the BC Archives, and Lissa Wadewitz shared material that she had gathered at the US National Archives. Barbara Lane, whose work on and for Native peoples in British Columbia has been an inspiration to many, shared some of that inspiration with me. The comments of Keith Carlson and two anonymous reviewers, and the copy-editing of Audrey McClellan, improved the manuscript substantially. Finally, Randy Schmidt and Darcy Cullen, editors at UBC Press, helped to steer this project from manuscript to book. This result is in many ways the product of the space that York and UBC, two marvellous public universities, have provided me. For this opportunity I thank the Canadian public and its commitment to post-secondary education.

I owe many thanks to the librarians and archivists who maintain the collections of historical documents and legal materials that I relied upon and who helped me find innumerable sources. I would particularly like to thank Anne Seymour at the Legal Surveys Division of Natural Resources Canada in Vancouver, whose detailed knowledge of the records of the Indian Reserve Commissions was invaluable; Balfour Halévy at the Osgoode Hall Law School Library; Anna Holeton, Mary Mitchell, and Sandra Wilkins at the UBC Faculty of Law Library; Ralph Stanton and George Brandak at UBC Library's Rare Books and Special Collections; the archivists at the British Columbia Archives; Bernice Chong at the Law Society of British Columbia; Elida Peers at the Sooke Region Museum; and Francis Mansbridge at the North Vancouver Archives and Museum. Lawyers Matthew Kirchner, Louise Mandell, Clo Ostrove, John Rich, and Harry Slade generously provided access to trial transcripts and appeal books. Trevor Chandler gave me lodging in Lillooet and pointed me in useful directions as I explored that area. Randy Bouchard and Dorothy Kennedy supplied me with material on the Squamish fishery and the Domanic Charlie case. Susan

Marsden confirmed some of my hunches and corrected others as I sought to understand the Tsimshian fishing leases. UBC students Katie Armitage, Shelley Balshine, Michael Begg, Ian Mosby, and Chris Wendell all provided able help as research assistants. Finally, Eric Leinberger in UBC's Department of Geography has again drawn a marvellous collection of maps that accompany the text.

I have also been privileged to join many conversations with First Nations people from across the province. In particular, Cliff Atleo Jr. spent a day with me, travelling by boat down the Alberni Canal through Nuu-chah-nulth territory, and Albert (Sonny) McHalsie led me on several occasions through Stó:lō territory in the upper Fraser Valley and lower Fraser Canyon. The Musqueam Nation, on whose traditional territory the University of British Columbia sits, provided me with several opportunities to present my work and to engage in discussion about the fisheries. Leona Sparrow has been particularly supportive. In addition, the First Nations Legal Studies Program at UBC attracts some wonderful students, and it has been my privilege to teach and learn from many of them, particularly Halie Bruce, Bruce Stadfeld, and Ardith Walkem.

I thank the various organizations that generously provided funding for this project, including York University, the Law Foundation of British Columbia, Coasts Under Stress (funded by the SSHRC, NSERC, and associated universities), UBC Faculty of Law, the British Columbia Heritage Trust, and the Osgoode Society for Canadian Legal History.

Finally, my parents have helped me in innumerable ways, but not more than by being marvellous exemplars of how to lead fulfilling and meaningful lives as university academics, parents, and now grandparents.

Candy Thomson has been with me throughout, and it is to her that this book is dedicated.

LANDING NATIVE FISHERIES

Introduction

On 21 October 1925, Domanic Charlie swung his gaff hook into the Capilano River where it runs through Squamish Indian Reserve No. 5 Capilano. The river was low after a dry summer, but with the fall rains the chum salmon, which had been schooling near its mouth, began to move into the river to spawn. With the gaff – a detachable iron hook with a connecting line at the end of a long pole – Charlie impaled a salmon and pulled it from the river. He hooked a second and hauled it ashore as well.

Nearly forty years old, Charlie was a Squamish hereditary chief who held the name of See-qawl-tuhn. He had been born on the south shore of English Bay and spent his life working in the local lumber industry, running log booms on the Serpentine River near the Canada-United States border with his half-brother, August Jack Khahtsahlano, and then on the Squamish River. In a pattern common for many Native people in the late nineteenth and early twentieth centuries, Charlie supplemented his wage work in the industrial resource economy by recourse to a traditional economy, including the harvesting of fish. In years when paid employment was scarce, the obverse might be true: sporadic work in the wage economy supplemented a livelihood largely derived from traditional patterns of resource procurement. In the emerging economy built over Charlie's lifetime around seasonal and cyclical resource extraction industries (fisheries, forestry, mining) and agriculture, elements of a more traditional economy provided an important source of food and, in some cases, income for many Native people.[1]

The Squamish are a Coast Salish people who claim as their traditional territory the ocean and the land surrounding what is now English Bay, Burrard Inlet, and Howe Sound, as well as the drainage basins of the Squamish and Cheakamus rivers (Figure 0.1). Their village sites looked out over these bodies of water, and, as was the case for other Coast Salish peoples, some of whom shared these spaces, their traditional cultures and economies were built around marine plants and animals, particularly fish.[2] The Capilano reserve includes one of those village sites. Allotted in the early 1860s, the reserve surrounds the last stretches of the Capilano River as it flows through a small flood plain and delta before emptying into the Pacific Ocean at the narrows to the inlet that forms the port of Vancouver. Identified in 1916 in the report of the Royal Commission on Indian Affairs for the Province of British Columbia as a "village site and fishing station," the Capilano reserve had been the first of more than twenty reserves allotted to the Squamish. These reserves provided the Squamish – about 400 community members in 1916 – a small land base in their traditional territory.

The Squamish had fished their traditional territory before newcomers, primarily of European and Asian descent, began to settle on the shores around Burrard Inlet in the mid-nineteenth century. This history of prior use could be understood as one of the sources of Charlie's right to fish in the Capilano River; he was exercising a Native or Aboriginal right to fish.[3] Canada's Department of Marine and Fisheries (Fisheries) located the legal basis of the Squamish fishery elsewhere. So far as the department was concerned, Charlie was fishing under an Indian food fishing permit that it had issued pursuant to the *Fisheries Act*. This permit allowed its holder to fish for food at times, in places, and with particular technologies that were closed to other fishers. However, the department imposed limits on the food fishery, and at the beginning of 1925 it had closed the Capilano to all fishing except angling in order to protect what it considered an important sport-fishing river. The Capilano was renowned for its steelhead, a sea-going trout and a favourite of sport fishers. Given its proximity to the growing city of Vancouver, the river was heavily fished. Some had advocated turning the river over to an angling club.[4] That had not happened, but in 1925 Fisheries increased its surveillance of the Native fisheries on the river. In October, the recently appointed Fisheries officer, Austin Spencer, an angler himself, watched Charlie pull the two chum salmon from the water. Later that day he charged Charlie under the *Fisheries Act* with catching fish in the Capilano River "by means other than angling."

The case of *Rex v. Charlie* appeared on the docket in the local police court the next week. Several weeks later, in a reserved judgment, the magistrate acquitted Charlie. The Department of Fisheries had not made out its case against him, and, more importantly as general precedent, the magistrate ruled that the department had no jurisdiction to regulate Indians fishing on Indian reserves. That responsibility lay with Canada's Department of Indian Affairs (Indian Affairs). Fisheries hired a senior lawyer, appealed to the county court, and within a month had its conviction. Although Indians held the exclusive right to fish on their reserves, the judge ruled

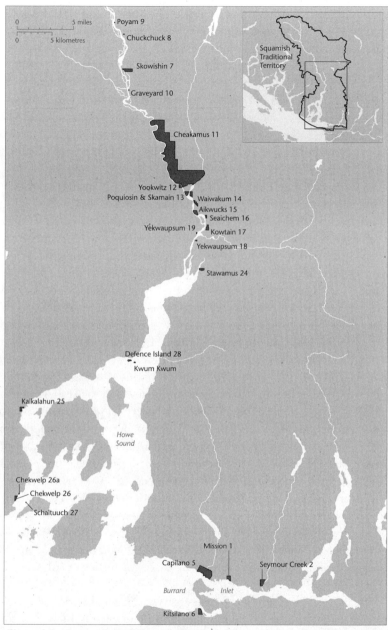

Figure 0.1 Squamish Indian reserves, 1925. The inset map shows the location of the reserves in relation to the area that the Squamish claim as their traditional territory. | *Source:* The boundaries of the traditional territory are based on maps submitted to and approved by the British Columbia Treaty Commission as the basis for treaty negotiation. (These maps are available at http://www.bctreaty.net/nations/soi_maps/Squamish_01_SOI_Map.pdf)

that Fisheries could regulate the Indian food fishery, even to the point of eliminating it to enhance a sport fishery. He fined Charlie one dollar. The Squamish wished to appeal, but Indian Affairs, which had also hired a senior lawyer for the appeal, had sparred long enough with its fellow department in the British Columbia courts. It refused to pursue the case or to release money it held in trust for the Squamish so that they could hire a lawyer to pursue it. The conviction stood. Charlie could not fish on the Capilano reserve, a reserve that had been allotted to the Squamish as a "village site and fishing station," unless he used a sport fisher's hook and line.

Charlie, a case that I return to in more detail in Chapter 6, marks the culminating effects of the processes examined in this book. By 1925, the governments of Canada and British Columbia had imposed on the Native peoples of the province an Indian reserve geography that presumed access to the fisheries. Over the same period, Canada had also constructed a legal regime governing the fisheries that, for the most part, opened them to all comers. An increasingly restricted and uncertain Indian food fishery was the only remnant of Native peoples' prior claim to the fish. These two legal constructs – the Indian reserve and the Indian food fishery – were two of the principal instruments of state power and colonial control in British Columbia. Although nested in two quite different legal regimes – one governing land use and structured around notions of private property, the other governing fisheries and premised primarily on the resource as common property – the reserve and the food fishery served the same purpose. Their intent and effect were to set aside fragments of traditional territories and fisheries for Native peoples, opening the remainder to immigrants. In short, the reserves and food fisheries were the colonial state's pinched concessions to the prior rights of Native peoples. Constructed together and operating in tandem, these legal categories consigned Native peoples to small parcels of land with inadequate protection for the fisheries that were to be their primary means of support. The story, then, is one of dispossession, a dispossession characterized by the colonial state's failure to honour its limited attempts to provide space for Native peoples and their livelihoods.

The connections between Indian reserves, Native fisheries, and Anglo-Canadian law in British Columbia need to be understood not only to explain the Indian reserve geography that remains largely intact today, but also to understand the impact on Native peoples of Canada's regulation of the fisheries. Although they are the subject of legal proceedings that have reached the Supreme Court of Canada, these connections have not been sufficiently explored or understood by the courts,[5] by scholars focusing on colonial land policy,[6] or by those writing on the fisheries.[7] The fisheries were absolutely central to the allotment of Indian reserves in British Columbia. Chronicling the connections between reserved land and the fisheries, and describing the legal regime that severed them, reallocating the fisheries to others, is the principal contribution of this book. Providing an understanding of the Indian food fishery – to the fisheries what the Indian reserve is to land – is another. These objectives are important not only for what they reveal about Indian land policy and the regulation of the fisheries in British Columbia, but also because the fisheries remain one of the principal sites of conflict between Native peoples and the state. Understanding the history of this conflict is central to its resolution.

The history of colonial land policy in British Columbia is now relatively clear. In *Making Native Space*, historical geographer Cole Harris has described the process of separating Native from non-Native land in the colony and province.[8] The provincial government's refusal to recognize Native title or to enter into treaties – with the exception of the Douglas Treaties on Vancouver Island in the 1850s – distinguished the construction of an Indian reserve geography in British Columbia from that of its neighbours to the east (the North-West Territories governed by Canada) and to the south (the Washington Territory of the United States).[9] The province had nothing to do with the negotiation of Treaty 8, one of a series known as the numbered treaties, that included the northeast corner of British Columbia. Over the vast majority of territory, immigrants settled on land for which Native title had not been ceded, a policy whose consequences reverberate today in a modern treaty process, in litigation over Aboriginal title,[10] and in negotiations over the government's responsibility to Native peoples when dealing with unceded or non-treaty land.[11]

Aside from the refusal to acknowledge title, Native land policies in British Columbia were also distinctive. Instead of detaching Native peoples from their traditional territories and placing them on large centralized reserves (a common pattern in the United States) or providing several substantial reserves within traditional territories (as was the pattern in much of Canada), the Dominion and provincial governments undertook a joint process in British Columbia that resulted in the allotment of many small reserves. They provided Native peoples with points of attachment within their traditional territories, but little more. In the 1920s, when the reserve allotment process came to an end, the acreage set aside as Indian reserve amounted to slightly more than one-third of one percent of the land area in the province.[12] As Harris has argued, echoing the voices of many Native peoples, the amount of reserved land was too small and its quality too poor to enable them to maintain viable economies in their traditional territories.

To the extent that Dominion and provincial officials sought to justify the unusually small reserve acreage in British Columbia, they did so on the grounds that Native peoples on the Pacific coast were primarily fishing peoples who did not need a large land base. Some agricultural and grazing land would be set aside as reserve, most of it marginal and often without sufficient water rights, but there was little enough viable farmland in the province, and immigrants would occupy most of it. Access to the fisheries was the principal basis on which government officials explained the land policy and on which Indian reserve commissioners allotted Indian reserves. To take one example from many, Indian Reserve Commissioner Peter O'Reilly described the Tlatlasikwala (Nahwitti) Reserve No. 1 Hope Island, at the northern tip of Vancouver Island, as "utterly worthless except as affording sheltered points from which the Indians can, weather permitting, start on their fishing expeditions."[13] In many parts of the province, the control of land – as a place to set a net or drag it ashore, as a rock from which to work a dip net, as a point of departure or return from fishing expeditions, or as a place to process fish – secured control of a fishery. This would change in the twentieth century as the widespread dissemination of gasoline-powered boats, refrigeration, and other technological changes created a much more mobile fishing fleet, but when Canada and British Columbia constructed an Indian reserve geography, the land/fish nexus mattered. Control of local fisheries provided a means, often the only means, of living on the land.

Government officials were right about the importance of the fisheries to Native peoples along the coast and through much of the interior. The Pacific Ocean provided an enormous abundance and diversity of plant and animal life that sustained large, geographically established, wealthy, and frequently hierarchical societies. Similarly, the river systems of the interior, which bore some of the world's largest salmon runs, were the oceanic tendrils along which Native societies and economies flourished. Lives were lived in seasonal cycles that included movement between a number of resource procurement and ceremonial sites, but most Native peoples in British Columbia were not hunter/gatherers as anthropologists have come to understand the designation. Instead, they were specialists in the harvesting and processing of fish, primarily salmon. Several scholars have written well about Native peoples' use and management of the fisheries along the Pacific coast of North America and the rivers that run to it, and I do not intend to add to that literature here.[14] Nevertheless, these were the conditions that led government officials to emphasize the importance of the fisheries and to discount the need for a large land base when they explained their land policy and justified the particular, and meagre, Indian reserve geography in British Columbia.

The characterization of Native peoples in British Columbia as fishing peoples, while largely accurate, was not neutral. The act of fishing was filled with cultural meaning in British society, a meaning that varied considerably depending on who was fishing and for what reasons. Fishing for pleasure was a guarded prerogative of the landed gentry in Britain, guarded not only to protect stocks but also to prevent the vices of idleness and indolence that the upper classes associated with fishing among the lower. Subsistence fishing was thought an activity of the mean and destitute, while commercial fishing deflected attention that was more appropriately engaged on the land.[15] A variant of this sentiment is captured in the writing of A.C. Anderson, a Hudson's Bay Company trader from 1832 to 1858 and later Canada's first inspector of fisheries in British Columbia and appointee to the Joint Indian Reserve Commission, who described the fishing peoples of the northwest coast of North America as follows: "Procuring an abundant livelihood with little exertion; gross, sensual, and for the most part cowardly – the races who depend entirely, or chiefly, on fishing, are immeasurably inferior to those tribes, who, with nerves and sinews braced by exercise, and minds comparatively ennobled by frequent excitement, live constantly amid war and the chase."[16] A little more than a decade later, Anderson was one of the early defenders of the Native fisheries, but his sense that fishing produced undesirable social consequences reflected broad cultural assumptions. They help to explain various efforts to deflect Native peoples from their fisheries and towards an agrarian economy. But the desire to leave what little agricultural land there was for incoming settlers, and the attachment of Native peoples to their fisheries, produced an Indian reserve geography based around the fisheries.

The maps in the Appendix reveal the locations of nearly 750 reserves across British Columbia that were identified specifically for their importance in the catching or processing of fish.[17] They show a particularly strong connection between reserved land and fisheries along the west coast of Vancouver Island and along the mainland coast to the Alaskan panhandle. In these areas, most of the reserves were intended as fishing stations that would provide access to a variety of fish, including salmon, halibut, herring, dogfish, cod, and oolichan, as well as clam beds and sea mammals, primarily seal. The single most productive fisheries, however, were in the interior at particular locations along the Fraser, Skeena, and Nass rivers. The anadromous Pacific salmon – primarily sockeye but also chinook, coho, chum, and pink – as well as locally important trout fisheries supported an extensive Native population with food and valuable trade commodities. As a result, many of the reserves in the interior were allotted to secure access to these resources. For both coast and interior, the maps represent a land-based human geography of the fisheries.[18]

Many more reserves could easily have been included on the maps. They show only those reserves that were expressly allotted by one of the reserve commissions for fishing purposes or for the processing of fish. Some reserves, including a few crucially important fishing sites, are not included because the reserve commissioners overlooked the importance of the fishery in allotting the reserve or did not record it. In some cases, the reserve was a village site, and this use is noted rather than the fishery. In other cases, the commissioners allotted land for wood or timber but did not mention that the wood might be used in smokehouses or to build fish boats and canoes. Finally, for a great many reserves the commissioners recorded fishing as one among several occupations of the residents but did not connect the reserve itself to a fishery. These reserves are not on the maps, not because the fisheries were unimportant, but because there is no indication in the reserve commissions' records that the reserve itself was allotted to secure access to a fishery. Indirectly, however, by providing a place to live and possibly a means of support during the off-season, these reserves also supported a fishery. In short, most reserves were set aside to secure Native fisheries. Land followed fish.

However, beginning in the 1870s, and coinciding with the rapid emergence of a canning industry, Canada's Department of Fisheries undertook to unravel the connection between land and fish. It sought to ensure that any proprietary interest held by Native peoples in their reserves remained on dry land; the exclusive rights that characterized the occupation of land, Fisheries maintained, did not extend to the fisheries. In fact, the Department of Fisheries based its opposition to the recognition of Native fishing rights on the grounds that while land, including Indian reserve land, might be held as private property, the fisheries were common property. The idea that fisheries were common property, which Dutch legal theorist Hugo Grotius gave voice to in international law and which the English common law gave expression to in the doctrine of the public right to fish, marks a fundamental difference between the law that applied to the land and that operating over water. Understanding these legal regimes and the differential extension of state sovereignty over land and water is crucial to understanding the processes of dispossession in British Columbia.

Sovereignty and Property – Land and Fisheries

Conflict over territory lies at the heart of colonialism.[19] This is most evident in settler colonies, such as British Columbia, where immigrants arrived with aspirations to build new lives and stay. Their settlement depended on the opening of territory that they could physically occupy and legally possess, and it was the function of the colonial state to open it. In some cases, negotiation and agreement with indigenous peoples underwrote the resettlement of territory by an immigrant population; in others it was military power and its violence that made the land available. In most places, these strategies of dispossession coexisted. In a mid-nineteenth-century colonial undertaking, such as British Columbia, military power lay in the past and in the background; its threat rather than its exercise was generally sufficient to prevent uprisings. So too the cultural assumptions of progress, superiority, and civilization buttressed the colonial state. But a growing literature highlights the importance of European law in the extension of imperial power overseas, both as a means of establishing control and as a justification for it.[20] State law established order, displacing its opposite, anarchy, and was a pivotal marker of progress, or so many assumed. In fact, the colonial theatre was a site of plural legal orders, indigenous, imposed, and hybrid, that coexisted, although not on equal terms. Where separate and formally recognized systems of customary law emerged, as in much of Africa and Asia, European law remained paramount in instances of conflict. In settler colonies, where the primary object was settlement of an immigrant population, colonial states did not construct separate legal systems. Instead, the salience of indigenous legal traditions withered as the balance of power shifted towards the immigrant society, the colonial state, and its legal structures.

Land

The opening of space to immigrant settlement involved two interrelated and concurrent processes that also defined the emergence of nation-states in Europe: the acquisition of sovereignty over a territory, and the consolidation of control within it. The acquisition of sovereignty involved establishing and defending a territorial claim within which the state held supreme law-making authority. The Peace of Westphalia (1648), which established a temporary reprieve from decades of conflict in Europe, is widely considered the moment when emerging nation-states established the principle that each was sovereign.[21] The basis of political authority had shifted away from a set of personal relationships between the sovereign and subject, and towards a notion of exclusive jurisdiction within defined territories.[22]

The consolidation of control within sovereign territory was part of a process that confirmed state sovereignty and was also an exercise of that sovereignty. Anthropologist James Scott describes the emerging state as the locus of centralized power in post-Enlightenment Europe, a development that, he argues, paralleled the efforts of these new political entities to reduce the complexities that characterized the customary rights and communal uses of feudal land tenure. Understood locally, the rights of land use in feudal society were inseparable from their context and opaque at a distance. A decisive moment in the emergence of the modern state, Scott suggests, was the simplification of land tenure, a process that enabled the central classification of land and, in turn, the extension of state power.[23] The state consolidated its sovereignty through the extension of its law of property.

The marginalization of custom was gradual and never complete. E.P. Thompson's marvellous studies of eighteenth-century rural England reveal a society riven by customary claims on the one hand, and rights based in the common law and statute on the other.[24] Custom did not disappear entirely – it remained a source of particular laws and, at another level, of the common law itself – but it had become subject to the jurisdiction of the common law and subordinated to it. As the diversity and salience of customary claims disappeared, the land was filled with a law of property that could be known at a national level. Legal texts such as Littleton's *Tenures* in the fifteenth century, Sir Edward Coke's *Institutes of the Law of England* in the sixteenth, and Sir William Blackstone's *Commentaries on the Laws of England* in the eighteenth made that law known.[25]

In a regime of private property, the owner's principal claim is to a right to exclude others from occupying, possessing, or otherwise using the thing claimed.[26] It is this "idea of ownership," suggests legal theorist Jeremy Waldron, that is the "organizing idea" of private property.[27] Blackstone famously described the right to private property as "that sole and despotic dominion which one man claims and exercises over the external things of the world, in total exclusion of the right of any other individual in the universe."[28] By the nineteenth century, this "ownership model," to borrow Joseph Singer's characterization,[29] had become the pervasive understanding of property in Britain and its settler colonies. The onus lay on non-owners to justify interfering with the right to exclusive possession. This had not always been the case, but by the mid-nineteenth century, private property was hegemonic, at least when people thought of land. Common property, characterized not by the right to exclude but by its opposite – the right not to be excluded – was consigned a secondary role in relation to land.[30]

The assertion of sovereignty and the imposition of a regime of private property occurred in British colonies overseas, albeit more suddenly and completely. In those colonies where colonization – the settlement of immigrant populations – was the principal imperial objective, the appropriation of territory and its reorganization assumed a particular urgency.[31] Here the doctrine of discovery was important in the European effort to establish spheres of sovereignty; imperial states that "discovered" a territory were entitled to take possession of it to the exclusion of other imperial powers – they were sovereign.[32] Sovereignty could also be acquired through treaty or war, and even by the act of settlement itself. In western North America, the Oregon Treaty of 1846 resolved the competing claims of Great Britain and the United States by dividing their interests along the forty-ninth parallel.[33] This treaty, dividing territory between imperial powers, would lead to the construction of a colonial state in British Columbia. It was part of the extension of imperial power, a process that historical geographer Daniel Clayton, in his study of the cartographic and geopolitical processes that placed Vancouver Island in a corner of the British imperial imaginary, has described as the "loss of locality."[34] The spaces of indigenous peoples became territorial possessions within empires. This was the legal scaffolding that would facilitate the emergence of a colonial state and, with it, a settler society.

Initially, the imperial state sought to interpose itself between indigenous peoples and settlers by acquiring land from its indigenous inhabitants – sometimes through purchase and treaty but at other times by war. The effect in law was to sweep away the complexity and, from the state's perspective, clutter of pre-existing customary rights and to create space devoid of legally recognizable tenure. The colonial state did not need to accommodate, in Scott's words, the "luxuriant variety of customary land tenure" in its efforts to standardize and centralize.[35] Instead, it deployed territorial strategies to empty spaces of indigenous tenure. Its sovereignty established and Native title ceded or ignored (as was the case in British Columbia), the colonial state could then refill the space with its land grants.[36]

In colonies of the British Crown, the recipients of a Crown grant received what had become the ubiquitous form of property – a fee-simple interest in free and common socage, the largest bundle of property rights known in English law.[37] But whatever its particular form, settlers acquired land through Crown grants of standardized bundles of rights. The effect was to bury a prior indigenous legal order, part of a much larger cultural assault that relegated Native peoples to the margins of settler society.[38] As anthropologist John Comaroff suggests, "it was by appeal to a specifically legal sensibility that the geography of colonies was mapped, transforming the land of others – typically seen by Europeans as wilderness before it was invested with their gaze – into territory and real estate; a process that made spaces into places to be possessed, ruled, improved, protected."[39] The idea of ownership – of private property – lay at the core of this sensibility.

The rights to private property, distributed and backed by the colonial state, conferred power on their holders who, by erecting a fence or otherwise enforcing a property line, gave local effect to the generalized act of dispossession inherent in the colonial encounter.[40] Private property was a form of state-delegated sovereignty over fixed parcels of land.[41] And given the virtues attached to private property in the heyday of mid- and late-nineteenth-century laissez-faire liberalism – for utilitarian philosopher Jeremy Bentham, the division of the English commons into private property was "one of the greatest and best understood improvements" – the extension of private property was itself a justification for imperial control.[42] The apogee of laissez-faire liberalism and of European colonial expansion coincided.

Indian reserves fit within this framework, although the delegation of sovereignty was incomplete. In British Columbia, as in the rest of Canada, the federal government held title to reserved land in trust for the Indian band to which it had been allotted. Together with the *Indian Act* of 1876, this arrangement was intended to facilitate the management of Native peoples and their integration into immigrant Canadian society.[43] Nonetheless, each band held the right of exclusive possession to its reserve land. Within the reserve, the band might distribute rights on the basis of hereditary entitlements, but this was difficult because small reserves excluded most of each band's traditional territory. Moreover, the state encouraged systems of property within reserves that mimicked the regime of fee simple beyond. Individual band members could apply for certificates of possession that established their right to defined parcels of reserve land.[44] Beyond the reserves, Native purchase or pre-emption of Crown lands became virtually impossible.[45] In terms of their capacity to hold property in land, the law confined Indians in British Columbia to their reserves, but the reserves were to belong to them exclusively.

Fisheries
The efforts to establish and consolidate the sovereignty of colonial states did not end at the foreshore. In colonies with a coastline or inland waterways, particularly those with valuable fisheries or strategic navigation routes, the colonial project included efforts to extend control over water and the resources within it.

The extension of state sovereignty over maritime territory proceeded incrementally. The consensus in the international community in the nineteenth and for much of the twentieth century was that state sovereignty over adjacent seas extended three miles from the shoreline. This distance, suggested by the Dutch legal scholar Cornelius van Bynkershoek in the early eighteenth century, was a compromise between those, such as legal theorist Hugo Grotius, who advocated the complete freedom of the seas, and others, such as English scholar John Selden, who argued that states were capable of extending their dominion and establishing property interests in the high seas. Bynkershoek posited that possession might "be regarded as extending just as far as it can be held in subjection to the mainland," and that this extended "as far as cannon will carry," a distance understood in the early eighteenth century to be approximately one league or three miles.[46]

This relatively small maritime belt of state sovereignty coincided with prevailing assumptions, held into the twentieth century, about the limitlessness of the ocean and the inexhaustibility of its resources. Beyond the three-mile limit, *mare liberum* (freedom of the seas) prevailed. In the words of Grotius, "the sea can in no way become the private property of any one, because nature not only allows but enjoins its common use."[47] The fish within the sea, therefore, were "exempt from such private ownership on account of their susceptibility to universal use; and as they belong to all they cannot be taken away from all by any one person any more than what is mine can be taken away from me by you."[48] Beyond the narrow maritime belt of state sovereignty, fisheries were common property – every state had the right not to be excluded.[49]

Within the maritime belt, the legal regime governing fisheries varied with the state, but the characterization of fisheries as common property was not easily dislodged. Under the English common law, the right to fish was understood to belong to the Crown. However, the Crown's ownership was subject to the public right to fish. In effect, the Crown held the fisheries in trust for the public. It could not, therefore, claim the exclusive right to fish or alienate that right to another party. Parliament could modify this rule, authorizing the Crown to grant exclusive fisheries, but the underlying presumption in the common law was that the Crown's subjects had a right not to be excluded from the fisheries – they were common property.[50]

In non-tidal rivers and lakes, sites of locally important fisheries, the legal regime followed the forms that applied to land. Under the English common law, rivers and lakes were subject to private ownership. In fact, the law created a presumption that property interests in land included the bed (the *solum*) of adjacent bodies of water to their midpoint and, therefore, the exclusive right to fish in those waters. The right to fish could be severed from the adjacent land, but the common law presumed otherwise. In short, the exclusive right to fish followed the right to exclusive possession of land. The courts modified this rule in North America, limiting the extension of private interests to the foreshore if the body of water adjacent to private land were navigable. The bed of navigable bodies of water in Canada belonged to the Crown, but it lay within the Crown's prerogative to alienate this interest, including the right to fish. In non-tidal waters, therefore, the extension of state sovereignty and the property regime that consolidated it emulated the regime on land.[51]

In tidal waters, sites of the most productive fisheries, the common-law right not to be excluded was also part of the state's consolidation of control over its sovereign territory. As with the grant of private property, the right not to be excluded emanated from the state and was assumed to supplant whatever preceded it. Even though the legal forms were different (in fact, they were opposites), the effect of their imposition was to erase pre-existing legal regimes, thereby confirming the sovereignty and consolidating the control of the colonial state. Except for the limited Indian food fishing privileges that the Department of Fisheries was prepared to concede, Native peoples had no prior claim to or property interest in the fisheries. The fisheries were common property in the sense that all subjects of the Crown, including Indians, were equally entitled to participate; everyone had the right not to be excluded.

However, just as Fisheries framed its approach to the Native fisheries in legal terms, so did its opponents. Native peoples maintained that they owned the fisheries that were inseparable parts of their communities and, therefore, that they had an Aboriginal right to fish, or that they had rights to fish that flowed from their Aboriginal title. Although this language is in some senses new, appearing since Aboriginal rights were entrenched in the Canadian constitution in 1982 and then interpreted in Canadian courts, the arguments are not. From the beginnings of a European presence in what is now British Columbia, Native peoples clearly articulated their rights to land and to resources, none more strongly than the right to their fisheries. This is reflected in the fisheries clause – the right to "fisheries as formerly" – in the Douglas Treaties, the fourteen agreements negotiated between Native peoples on Vancouver Island and the Hudson's Bay Company as representative of the British Crown. This treaty right became a touchstone for those who sought to find space for Native peoples' fisheries, and I seek to understand it, in conjunction with the emerging colonial land policy, in Chapter 1.

After the Douglas Treaties, the work of the Indian reserve commissioners provides the clearest formal recognition of the importance of the fisheries to Native peoples. In Chapters 2 to 5, I turn to the years between 1876 and 1910 when the Dominion and provincial governments constructed and then implemented a land policy that was explicitly premised on access to the fisheries. Chapter 2 covers the work of the Joint Indian Reserve Commission (1876-78) and that of Gilbert Malcolm Sproat (1878-80) as the sole commissioner of the Indian Reserve Commission. Sproat's work is characterized by careful and thoughtful, if largely futile, attempts to protect the Native fisheries as a means of finding sufficient space for the economies and cultures of Native peoples and immigrants.[52] In Chapter 3, I consider the first year of Sproat's successor, Peter O'Reilly, the effective and efficient colonial administrator who, more than any other, would implement the province's vision of an appropriate Indian land policy in British Columbia. It is surprising, therefore, that it was the work of this commissioner, inclined to follow instructions and not to advocate, as his predecessor had done, that most provoked the Department of Fisheries. On his first circuit to the middle Fraser and then to the north coast, O'Reilly granted exclusive fisheries along the Fraser and Nass rivers as part of or in addition to the reserves grants, something the department adamantly refused to recognize on the grounds that the grants violated the common-law doctrine of the public right to fish. In Chapter 4, I explore the public right to fish, its interpretation in Canadian courts, and its role in shaping Native peoples' access to the fisheries. I then return, in Chapter 5, to O'Reilly's continuing work as reserve commissioner over nearly two decades, and to the work of his successor, A.W. Vowell, from 1898 to 1910.

In Chapters 6 and 7, I detour from the narrative of reserve creation to describe the statute-based regulation of the fishery and its impact on Native fisheries and fishers. It is in Chapter 6 that I revisit *Charlie* as part of the construction, in law, of an Indian food fishery. In detaching Native peoples from their fisheries, the state also defined the terms under which they could participate as workers in the industrial commercial fishery. As Alicja Muszynski has argued, Native participation was gendered – men fished, women worked in the canneries – but it was also racialized.[53] By this she means that the canneries used gender and race as markers to define and limit who might work, in what capacity, and for what remuneration. Native fishers and cannery workers were not the only ones to bear these markers. Japanese fishers and Chinese cannery workers were also the targets of discriminatory attempts to limit or end their participation in the industry. However, in Chapter 7 I focus on the opportunities for Native fishers in the commercial fleet and on the ways in which the allocation of fishing licences worked to preclude full Native participation in the industry.

I return to the allotment of Indian reserves in Chapter 8 to consider the importance of the fisheries in the work of the Royal Commission on Indian Affairs for the Province of British Columbia – the McKenna-McBride Commission – and in the final agreements between the Dominion and province that wound down the reserve process, fixing the reserve geography of British Columbia that largely remains to this day. Although the commission was to focus on the land question, the commissioners found themselves repeatedly drawn by compelling Native testimony to the fisheries, an issue they could not avoid because the capacity of the reserved land to support viable livelihoods depended, by design, on access to fish.

These chapters are full of place names and discussions of particular locales. This is in part a function of the fisheries. Fishing is not equally good everywhere. Prized locations depend upon innumerable physical and ecological variations (in bodies of water, climate, and species of fish), but are also determined by the prevalent fishing and processing technology and the social/cultural/legal milieu in which they are used. When the reserve commissioners sought input from Native peoples, they heard general statements of ownership, but also specific requests for control of particular sites that were essential parts of local Native economies. More often than not, these were fishing sites. Reflecting this attention to place, the chapters and the appendix are full of maps that, I hope, help the reader make sense of detailed descriptions and identify larger patterns. They also provide a sense of the spatial theatre that law constructs and operates within.

The focus of this book follows from its sources, principally the records of the Indian reserve commissions, the Department of Indian Affairs, and the Department of Marine and Fisheries. These were the authorities most involved in creating Indian reserves and in regulating the fisheries. I spent time with the provincial Department of Fisheries records, but they appear infrequently here. In its early years, most of the provincial department's efforts went into enhancement projects rather than the regulation and management of the fisheries. Other material that I have dipped into includes cannery records and private manuscripts from those involved in the fishing industry. Had the study paid as much attention to management of land as it does to fish, the files of the provincial Department of Lands would have been indispensable. But taking this book on its own terms – as a study of the connections between the colonial land policy and the regulation of the fisheries – what is most glaringly absent are the Native voices. To the extent that they appear, sympathetic accounts are almost always filtered through the ear and pen of a missionary or, more commonly by the late nineteenth century, an Indian agent or reserve commissioner. More antagonistic representations of Native claims come from the Fisheries officers. It is primarily through the transcripts of the hearings conducted by the Royal Commission on Indian Affairs for the Province of British Columbia, 1913-16, that Native voices appear in the written record in their least mediated form. Here one finds Native voices that eloquently, sadly, and sometimes angrily denounce the laws and policies of the colonial state that refused to recognize the legitimacy of their claims and consigned a great many to lives of poverty. One also catches glimpses of a suppressed legal order struggling to survive the hegemony of the colonial legal order. Beyond the written records, which can only be the most partial representation of Native views, are the oral histories that reside in Native communities. Ethno-historian Keith Carlson has skilfully used those histories and the written record to describe the roots of a continuing conflict between First Nations over the fisheries at Yale.[54] In this book I deal with the laws and policies of the colonial state, their contradictions and ambiguities, but also their power. To that end, I have attempted to map officially recognized fishing sites, describe the conflicts that ensue, and explore the role of law in the process. I have not attempted the larger and more difficult challenge of mapping the patterns of Native peoples' use and control of the fisheries – in effect, the legal geography of the Native fisheries. Much of that work is being undertaken in First Nations treaty offices around the province and may well result in the most fundamental re-mapping of the territory since the reserves were created in the late nineteenth and early twentieth centuries.[55]

1
Treaties, Reserves, and Fisheries Law

The west coast of North America entered the imperial orbit of Britain in the late eighteenth century when maritime explorers mapped the coast, and merchants followed to trade with Native peoples for sea otter pelts. The coastal trade was vigorous, but seasonal and short lived. In the early nineteenth century, other traders arrived from the east, linking the territory overland to Hudson's Bay, the Red River Settlement, and Montreal, and to markets around the Pacific. Trading posts became the first non-Native settlements in the western cordillera. Personnel rotated through the posts, but the trading companies – until 1821 the North West Company and then the Hudson's Bay Company (HBC) – remained, their corporate presence a constant despite the comings and goings of individual traders and workers. Native peoples responded to these new sources of wealth and power by, in some cases, reorganizing their economies to take advantage of the window into world markets that the trading companies provided or, in other cases, resisting and retreating. In either event, Native peoples continued to live in territories they still largely controlled. Apart from the sites of their forts, gardens, a few farms, and the temporary spaces that their workers occupied as they moved between posts, the trading companies did not seek to control land.[1]

It was not until 1849, after the Oregon Treaty of 1846 extended the forty-ninth parallel as the boundary between American and British interests in North America to the Pacific, that the HBC assumed responsibility for Vancouver Island as a proprietary colony of the British Crown with a mandate to encourage settlement.[2] Settlers, primarily of British descent, and many of them retired HBC employees, trickled into this distant corner of Empire until 1858, when, with the discovery of gold in the Fraser Canyon, thousands of miners flooded north from the gold fields in California to Victoria, stopping briefly for supplies before crossing to the Fraser River on the mainland. To forestall a possible annexation of the territory by the

States, and to establish British law, if only symbolically, over a well-
and defiant collection of miners, the Colonial Office created the
id colony of British Columbia that same year. It also sent a detach-
rom the Royal Engineers, 220 soldiers strong, that for the next five
rovided a land-based military presence to complement the gunboats
Royal Navy and that began building a colonial infrastructure, in-
ng roads, bridges, townsite plans, and land surveys. The island and
land colonies joined in 1866 and entered the Canadian confederation
871.[3]

The mid-nineteenth-century transition in British interest in the western
cordillera – from trading country to settlement frontier – also marked the
beginnings of a fundamental transition in relations with Native peoples.
Associates in the diversified trading enterprise that had emerged in the first
half of the nineteenth century, Native peoples now became obstacles in the
way of settlement and progress.[4] During the initial and brutal encounter of
miners and Natives in the Fraser Canyon in 1858, miners shot their way in
and used their superior firepower to hold their positions.[5] Less suddenly,
but ultimately with greater implications for the lives of most Native peoples
in the two colonies, the fixity of an agrarian economy and accompanying
notions of private property clashed with the seasonal patterns of Native
land use that revolved around winter villages, fishing and hunting grounds,
plant harvesting territories, and spiritual sites. The appearance of a settler
fence to satisfy the requirement under a pre-emption claim to "improve"
the land, to keep foraging animals out of a vegetable garden, or simply to
establish a tangible boundary enclosing private land, was the locally encountered evidence of a new and imposed system of private property. Land
inside the fence had become the exclusive possession of its owner.[6]

Supporting the fence and the right to exclusive possession that it an-
nounced were cadastral surveys, maps, land registries, courts, police, and
the military – the technologies and institutions deployed by the state to
extend and confirm its power by defending the inviolability of the rights it
had granted. Indeed, one of the principal functions of the colonial state
was to protect these new property lines, securing private property and creating transferable value for its owner. Geographer Nick Blomley suggests
that the frontier, which marked the spatial limit of the colonial state's effective control, was distinguishable by the different systems of land tenure
that existed on either side: "Inside the frontier lie secure tenure, fee-simple
ownership, and state-guaranteed rights to property. Outside lie uncertain
and undeveloped entitlements, communal claims, and the absence of state
guarantees to property."[7] Similarly, historian John Weaver describes "bor-
derlands or frontiers as areas where the colonizer's regime of property rights

had not been firmly installed, but where newcomers were already marking out places in anticipation of that condition."[8] Within the colonial state, life was civilized (propertied); beyond, it was savage (not or insufficiently propertied).

Houses, gardens, and fences were the markers of possession within English culture, establishing not only individual ownership, but also evidence of British sovereignty, civilization, and, in relation to Native peoples, superiority. They were the signs, to the culture that produced them, of an industrious people. The importance of these domestic and mundane acts of building and planting as symbols of British dominion separated the British not only from the Native peoples, who lived differently in the same territories, but also from other European colonial powers in the western hemisphere who, Patricia Seed argues, based their claims to sovereignty on other ceremonies of possession.[9] This assertion of sovereignty and the regime of property that accompanied it displaced prior ways of living on the land. Blomley emphasizes the violence that underscored the dismembering of a locally owned and regulated commons, the creation of private property and its right of exclusive possession pushing out those who once had access.[10]

As the Crown began to alienate parcels of land to newcomers through processes of pre-emption (a claim to land that could be perfected once the land was surveyed, the purchase price paid, and evidence of improvements confirmed) or purchase, it also allotted Indian reserves. Beginning with the creation of the Colony of Vancouver Island in 1849, these processes, which divided land into Indian reserves and land available for the exclusive possession of immigrants, continued in tandem for seventy-five years, the allotment of reserves generally following a few years behind increased settler interest in a region, until 1924, when the Dominion and provincial government agreed on what they regarded as the final reserve geography of British Columbia. With the exception of the Douglas Treaties on Vancouver Island and Treaty 8 in the northeast, the issue of Native title and the need to seek agreement with Native peoples over the shared use of space was ignored.

In this chapter, I begin a parallel narrative of the conflict over fish, focusing on the process of reserve allotments and the introduction of Anglo-Canadian fisheries law. It starts with the Douglas Treaties and, more generally, the nature of Native title and of rights to fish. It then turns to consider the allotment of reserves in relation to the fisheries. This discussion is relatively brief, a function of the fact that fisheries were not yet a site of conflict between Natives and newcomers. The inexhaustible abundance of fish was presumed and, although some imagined the resource could be a future source of great wealth, there was little non-Native interest. That was not the case in the Great Lakes fisheries, already the site of protracted conflict between Native and non-Native fishers and the colonial government in Upper Canada and then Canada West. It was this history of conflict that British Columbia acquired when it joined the Canadian confederation in 1871 and ceded jurisdiction over fisheries to the Dominion. The chapter concludes with a brief foray into that history.

Treaty and Native Rights to Fish

The agreements known as the Douglas Treaties are fourteen land purchases made by James Douglas in his capacity as the Hudson's Bay Company's chief trader, and then governor of the colony of Vancouver Island, between 1850 and 1854. The land, purchased from Native peoples on Vancouver Island, covered a small fraction of the island, including the area around Victoria, the Saanich Peninsula, the future townsite of Nanaimo midway up the island, and an area near Fort Rupert at its northeastern end (Figure 1.1). The treaties loom over the process of reserve allotments in British Columbia, marking the beginning of an unfinished project to treat with Native peoples and serving as a reminder of the suppressed yet outstanding question of Native title.

Much has been written about why Douglas undertook these purchases on Vancouver Island and why he did not continue them. It seems that recognition of a legal requirement to extinguish Native title was an important part of his motivation for beginning the process, but ebbing enthusiasm for treaties in the Colonial Office in London reduced the incentive to continue the process when other interests intervened. Cole Harris has emphasized Douglas's pragmatism, born of a lifetime in the fur trade, suggesting that he was less concerned about theories of Indian land policy and even of the law of Native title than about finding workable solutions for Native and European coexistence.[11] Legal historians Hamar Foster and Alan Grove suggest that the decision of an Oregon court to deny the existence of Native title, discredited in Oregon and Washington but picked up in Alaska, may also have influenced Douglas and his successor in the formation of colonial land policy, Commissioner of Lands Joseph Trutch, who was openly hostile to the idea of Native title.[12]

The legal standing of Native title may have been fragile enough in the mid-nineteenth century that colonial authorities were prepared to ignore it, but there was less doubt about the existence of specific Native rights, particularly rights to hunt and fish. Moreover, protecting these rights, on which Native economies depended, fit Douglas's pragmatism. Native peoples' hunting could coexist with non-Native ownership, if not use and

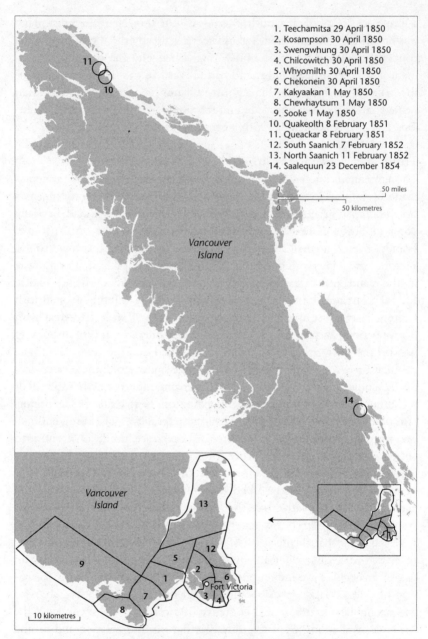

1. Teechamitsa 29 April 1850
2. Kosampson 30 April 1850
3. Swengwhung 30 April 1850
4. Chilcowitch 30 April 1850
5. Whyomilth 30 April 1850
6. Chekonein 30 April 1850
7. Kakyaakan 1 May 1850
8. Chewhaytsum 1 May 1850
9. Sooke 1 May 1850
10. Quakeolth 8 February 1851
11. Queackar 8 February 1851
12. South Saanich 7 February 1852
13. North Saanich 11 February 1852
14. Saalequun 23 December 1854

Figure 1.1   Boundaries of the Douglas Treaties, 1850-54. The treaty process did not continue beyond 1854, leaving the issue of Native title unresolved on the rest of Vancouver Island and throughout most of the mainland colony of British Columbia. | *Source:* The treaty boundaries were adapted from the information available in the Government of Canada's Directory of Federal Real Property (http://www.tbssct. gc.ca/dfrp-rbif/treaty-traite.asp?Language=EN).

occupation, of the land, and the fishery could be secured without much impact on the land available for incoming settlers. In anticipation of the treaties, Douglas wrote to the HBC that he "would strongly recommend, equally as a matter of justice, and from regard to the future peace of the colony, that the Indians Fishere's [sic], Village Sitis [sic] and Fields, should be reserved for their benifit [sic] and fully secured to them by law."[13] HBC secretary Archibald Barclay, in setting out the Company's obligations and policy towards Native peoples on Vancouver Island, instructed Douglas that the "right of fishing and hunting will be continued to them."[14]

On the basis of these instructions, Douglas entered negotiations with the tribes on southern Vancouver Island. After minimal discussions (of which no minutes were kept), Douglas asked the chiefs to place X's on blank sheets of paper. Following the conclusion of the first nine agreements at Fort Victoria between 29 April and 1 May 1850, Douglas wrote to the HBC to explain his understanding of what had transpired: "I informed the natives that they would not be disturbed in the possession of their Village sites and enclosed fields, which are of small extent, and that they were at liberty to hunt over unoccupied lands, and *to carry on their fisheries with the same freedom as when they were the sole occupants of the country*."[15] He forwarded the "signatures" of the chiefs and asked that the HBC supply the proper conveyancing instrument to which the signatures could be attached. Several months later, Barclay replied, approving the agreements and sending a template purchase agreement, based on New Zealand precedents, that would become the text of the Douglas Treaties.[16] The first paragraph described the lands that were covered by the treaty; the second described the terms:

> The condition of or understanding of this sale is this, that our [Indian] village sites and enclosed fields are to be kept for our own use, for the use of our children, and for those who may follow after us; and the land shall be properly surveyed, hereafter. It is understood, however, that the land itself becomes the entire property of the white people for ever; it is also understood that we are at liberty to hunt over the unoccupied lands, and *to carry on our fisheries as formerly*.[17]

Although the structure and content of Barclay's template emulated the New Zealand deeds, the final clause setting out the hunting and fishing rights was new. The guarantee that the Indians were to be "at liberty ... to carry on [their] fisheries as formerly" appears to be an abbreviated version of the agreement as described by Douglas several months earlier: "They were at liberty ... to carry on their fisheries with the same freedom as when they were the sole occupants of the country."

Given these events, the treaties are best understood as oral agreements. The written text, based on imperial precedent, drafted by someone not present at the negotiations, and supplied months afterward, should be considered as evidence of the terms of those agreements, not as the agreements themselves. As evidence, the written text probably provides reasonable indication of what the HBC thought it needed to do and how Douglas understood the treaties. The anthropologist Wilson Duff considered the text to be "the white man's conception (or at least his rationalization) of the situation as it was and of the transaction that took place."[18] It provides little or highly qualified evidence, at best, of how the Native participants understood the agreements.[19]

Even the terms of the written text are not self-evident.[20] It is clear, however, that "fisheries" were an important part of the agreement. A "fishery" or its plural, "fisheries," refers not only to the act of fishing but also to the places where it occurs. In reserving "fisheries," therefore, the Douglas Treaties reserved the right to fish at the places where Native people fished. Several years after concluding the last of the treaties, Douglas informed the Vancouver Island House of Assembly, in similarly broad terms, that Native peoples "were to be protected in their original right of fishing on the coast and in the bays of the Colony."[21] In describing the fishing right as "original," Douglas meant that it preceded the British assertion of sovereignty, not that it was otherwise constrained.

In short, the Douglas Treaties provided broad protection for Native fisheries. In the 1850s, the boundaries of the right did not need to be carefully drawn. An abundance of fish was presumed, and there was little non-Native interest in prosecuting a fishery. However, the fisheries were certainly not an afterthought. The HBC had deployed some of its workers to the fisheries of the Fraser River in the 1840s but had realized that it was more efficient and effective to purchase fish from Native fishers. These fish, which the HBC barrelled and salted on the Fraser beginning in the 1820s, had become one of its principal exports from the Pacific coast of North America.[22] Thus, the treaties were concluded in a context of well-established and ongoing commercial activity in the fishery involving the HBC and Native peoples. Douglas believed that this would continue, and he hoped that it would grow. It is hard to imagine, therefore, that the right to "fisheries as formerly" did not include a commercial aspect, such as the right to sell fish to commercial trading companies. Furthermore, there is no indication that Douglas thought that the treaty protected only a food fishery. In fact, viewing "Indian food fishing" as a separate category was not yet a way of thinking about Native fishing in British Columbia. The concept, established in Canadian fisheries regulations in the late nineteenth century, would become an important part of fisheries management and an effective way to diminish Native peoples' access to the fish, but it was not part of the framework in which the treaties were negotiated.[23]

However, Douglas certainly did not intend to preclude non-Native participation. He believed that the long-term prosperity of the colony depended on attracting immigrants, and the fisheries would be one of the principal draws for those newcomers. The HBC had sought control of the fisheries as part of the Crown grant of Vancouver Island, but the Crown withdrew this provision, which had appeared in an early draft, in the midst of public disapprobation of the HBC in London.[24] As a result, the HBC prospectus for the colonization of Vancouver Island informed prospective settlers that "*every freeholder shall enjoy the right of fishing* all sorts of fish in the seas, bays, and inlets of, or surrounding, the said Island."[25] In tidal waters, then, the prospectus asserted the right of the landowning public to fish as, indeed, the common-law doctrine of the public right to fish established for the public at large.

It was not until the creation and expansion of the industrial commercial fishery in the 1870s that Native rights to fish began to be challenged and that the meaning of the fisheries clause in the Douglas Treaties began to matter – and to be forgotten or ignored. In May 1878, complaints that the Esquimalt people were wasting fish roe were registered in the provincial legislature. Indian Reserve Commissioner Gilbert Malcolm Sproat noted that the allegations, if true, were to be regretted, but that the Esquimalt were a party to one of the Douglas Treaties, which protected their right to fish, and therefore the government could not interfere.[26] A settler's preemption at the mouth of the Goldstream River in Saanich Inlet was another source of concern because, as Sproat pointed out, it would interfere with the treaty fishing rights of several different bands that occupied the location seasonally.[27] Sproat, at least, interpreted the treaty right broadly.

What about Native peoples who were not party to the Douglas Treaties? The language of the fisheries clause – that Native people were "at liberty ... to carry on ... fisheries as formerly" – suggests that the treaties should not be understood as creating or granting a right to fish. Instead, the clause turned an existing practice and right into a treaty right. Native peoples in the rest of the province did not have this treaty right, but they still had rights that pre-existed the treaties. In 1860, Douglas wrote to the Colonial Office to describe a series of meetings with Native peoples in the interior of the mainland colony. Douglas explained that he had told the people gathered at Lillooet that "they might freely exercise and enjoy the rights of fishing the Lakes and Rivers, and of hunting over all unoccupied Crown Lands in the Colony."[28] Although clearly echoing the language in the treaties, the characterization of the rights to hunt and fish was somewhat narrower. The right to hunt extended only to "unoccupied *Crown* lands" and, without any reference to prior rights or to "fisheries as formerly," the promise that "they might freely exercise and enjoy the rights of fishing" was little more than what Douglas would have told a non-Native audience. The end of the treaties marked the end of Douglas's formal recognition of Native title, and perhaps, by 1860, he was being more circumspect in his recognition of rights to hunt and fish as well.

However, the fishing rights in the Douglas Treaties remained a powerful presence in the discussion of fishing rights beyond the borders of the treaties. Sproat was involved again in 1878 when the location of a sawmill became an issue because the running of logs down the Cowichan River to the mill threatened to destroy the Cowichan's weir fishery. He argued that the Cowichan, although not party to a treaty, had a similar right, by virtue of their long use of the river, to continue fishing as formerly. The government, he thought, should provide compensation and obtain the Cowichan's consent before the mill owner could float logs down the river.[29] The following year, A.C. Anderson, the senior Department of Marine and Fisheries (Fisheries) official in British Columbia and former member of the Joint Indian Reserve Commission, referred the minister of Fisheries to the Douglas Treaties, indicating his understanding that Native peoples across the province, not just the treaty Indians, had a right to continue their fisheries.[30] Four decades later, in 1918, William Sloan, the provincial commissioner of Fisheries, expressed the view that Native peoples had rights to their fisheries, that the fisheries clause in the Douglas Treaties was evidence of this, and that if Native fisheries were to be closed, even for conservation purposes, then the fishers should be compensated. As the devastating impact of the 1914 rock slide at Hells Gate on the Fraser River sockeye became apparent, he wrote:

> The runs of salmon to the spawning-beds of the Fraser have become so alarmingly attenuated that drastic measures will have to be taken to restore the runs. The measures to be taken must not only include the secession of all fishing in tidal limits for a period of years, but must be made to include all fishing above tidal limits by Indians for all time, notwithstanding that *they have both a natural and a treaty right* to take such salmon as they desire for food so long as they confine themselves to the gear originally used by them ...
>
> *The right of the Indians to take salmon is unquestioned,* but the number of salmon they can now catch is so small as to be of little benefit to them. Owing to the fact that most of the Indians now grow the bulk of the food they use and are no longer dependent on salmon, and that drastic measures must be taken to restore the salmon

of the Fraser, the Government should step in and acquire by purchase the Indians' right to take fish above the commercial boundaries. It is suggested that the Indians, if deliberately approached, would dispose of their fishing rights to the Government, and that the Government is fully warranted in entering upon negotiations to acquire those rights. The sooner the better.[31]

This understanding was certainly not unanimous. The Department of Fisheries considered Native fishing privileges – it did not consider them as rights – were derived from the Crown at its pleasure. More particularly, they derived, under legislation, from the department itself. This view came to predominate in the 1880s and 1890s, although the Department of Indian Affairs was never entirely comfortable with it, and there were always voices within government, such as Sproat's and, later, Sloan's, which insisted that Native fishing rights needed to be recognized and pointed to the Douglas Treaties as evidence.[32] Native voices were unequivocal, if seldom heard in the halls of the Department of Fisheries in Ottawa. Their fisheries were not a privilege, nor were they derived from Crown grant. They had rights to fish – rights that originated in their laws and legal traditions – that they had never surrendered.[33]

Fisheries and Colonial Land Policy

Cole Harris and Lillian Ford document the creation of 140 reserves on Vancouver Island and in the mainland colony of British Columbia between 1849 and 1871.[34] Of the twenty-eight reserves allotted on Vancouver Island before Confederation, most were Douglas Treaty reserves. Harris describes these allotments as forming the beginnings of a Native land policy that "focused on small reserves tucked within the cadastral survey of colonial settlement ... It was an imposed policy, one that took into some account the location of occupied winter villages, but that shows no evidence of meaningful consultation with Native people. It did provide, however, some minimal space for Native peoples within their traditional territories, a pattern that would endure."[35] The non-treaty reserves on Vancouver Island, most of which were between Duncan and Nanaimo, where growing non-Native settlement was causing considerable unrest among the Cowichan and neighbouring groups, followed a similar pattern.[36]

Although reserves were small, Native people were not confined to them. The hunting and fishing clause in the treaties protected their rights to continue these activities in traditional territories beyond the reserves, and it would have been assumed that these terms applied to the non-treaty groups as well. Probably because of these general guarantees in the treaties and because there was scant non-Native interest in the fisheries, there is little suggestion in the official record that the pre-Confederation reserves on Vancouver Island were allotted to secure access to fish, or that officials attempted to justify the small reserves on the grounds that fishing peoples did not need large land bases. That justification would come later. The location of the principal fisheries for some of these groups may also have been a factor reducing the likelihood of explicit connections between reserves and fisheries. The Songhees, whose reserves were in and around Victoria, fished sockeye with reef nets in the San Juan Islands. Their two reserves on Discovery Island and the Chatham Islands, located just off the southern tip of Vancouver Island, may well have been set aside as departure points for those fisheries across Haro Strait, but the fisheries themselves were in United States waters (Figure 1.2).[37] Although its effect was not immediately felt, the international boundary, which severed Native territories, would cause great hardship.[38] Dave Elliot Sr., a member of the Saanich, who also had fishing grounds across Haro Strait, recounted the impact of the border on his people and their access to the fisheries:

> We had all those salmon runs and that beautiful way of fishing [reef nets]. When they divided up the country we lost most of our territory. It is now in the State of Washington. They said we would be able to go back and forth when they laid down the boundary, they said it wouldn't make any difference to the Indians. They said that it wouldn't affect us Indians.
>
> They didn't keep that promise very long; Washington made laws over the Federal laws, British Columbia made laws over those Federal laws too, and pretty soon we weren't able to go there and fish. Some of our people were arrested for going over there. That's what happened to our fishery. That's why we're not fishing right now today. The Indians are fishing in Washington using our way of fishing. We lost our fishery, and our fishing grounds.[39]

Other groups along the east shore of Vancouver Island, particularly the Cowichan, Chemainus, Penelakut, and Halalt, had important summer fisheries on the mainland near the mouth of the Fraser River. They did not cross an international border to access their fisheries, but government officials did not allot reserves at these summer fishing camps either. This became a problem in the 1870s with the rise of the industrial/commercial fishery and the growing competition for fish.

On the mainland, there were no treaties between 1858 and 1871, but Ford and Harris count 112 reserves allotted primarily in the Fraser Valley and along the Fraser and Thompson river systems. It was here that settler interest in using Native land, largely for agriculture, mining, and road building, had produced the greatest tension and conflict. The records of exactly

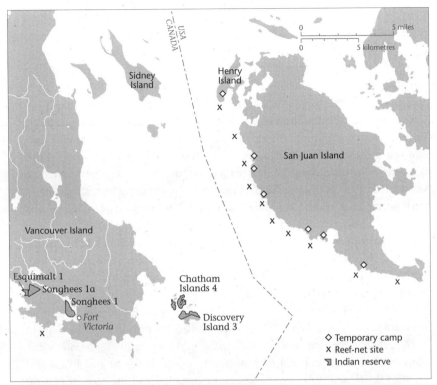

Figure 1.2 Songhees Indian reserves and fisheries. The reserves on Discovery Island and the Chatham Islands may well have been intended as points of departure for the reef-net fisheries on San Juan Island. However, the international boundary, established by the Oregon Treaty in 1846, would eventually prevent Songhees' access to those fisheries. | *Source:* The locations of the camps and reef-net sites are from John Lutz, "Work, Wages and Welfare in Aboriginal-Non-Aboriginal Relations, British Columbia, 1849-1970" (PhD diss., Department of History, University of Ottawa, 1994), 149, 224.

what was allotted in these non-treaty reserves are poor, not only because they do not survive, but also because the government did not record its work carefully. In some cases there were multiple and inaccurate surveys; in other cases there were no surveys at all. The problem of determining reserve allotments is compounded by the significant change in land policy that followed Governor Douglas's retirement in 1864. In comparison to what would follow, the Douglas reserves on the mainland were generous and, when combined with land laws that allowed Natives to acquire other land on the same terms as immigrants, seemed to provide space for Native people to participate as full members in the colonial economy. At least that was Douglas's goal. The new commissioner of Lands and Works, Joseph

Trutch, who replaced Douglas as the principal architect of a Native land policy, set about reducing the size of many of the existing reserves and erecting legal boundaries to limit Native acquisition of other land. His aim, shared by many in the colonial society, was to create as much space as possible for an incoming settler society. It was Douglas's views that were the more unusual.[40]

The reserves allotted under Trutch's tenure were small, to amount to no more than ten acres per family. This was insufficient to support viable agricultural economies, but agriculture was not the focal point of most Native economies in the province. The fisheries were far more important, and many of these small reserves were situated at important fishing sites. This is particularly true of the reserves allotted to the Stó:lō and Nlha7kapmx along the Fraser River. Fish, primarily salmon, were the principal source of sustenance and wealth for these communities, and the placement and distribution of reserves almost exclusively along the river reflects that connection. The fishery at Yale at the start of the Fraser Canyon, for example, was particularly productive, attracting thousands of fishers from downriver in July and August. Upriver there were a great many locally important fishing sites, some of which were included within the reserves. In the flood plain of the Fraser Valley, most of the reserves would have been allotted for some combination of agricultural and fishing purposes. It was at Kamloops and in the Okanagan where the connection between the pre-Confederation reserves and the fisheries was perhaps most tenuous. Fish were still an important part of local economies, but the large reserves, subsequently reduced, were intended to support a farming and ranching economy.

The colonial government after Douglas would likely have reduced Native access to fish, just as it reduced reserves, had fisheries been an important part of the colonial economy. They would become so, but only after British Columbia joined Canada and ceded jurisdiction over fisheries to the Dominion. Until then, the fisheries did not receive much attention. Reflecting both a broadly held perception in Britain that seacoast fisheries were best managed by the market's invisible hand and the relative lack of importance of the fishery to the colonial economy in British Columbia, the local colonial government hardly intervened in the fishery. The House of Assembly on Vancouver Island passed one act in 1862 that contained a clause purporting to regulate fishing in Victoria's inner harbour and the colony's lakes, but no one seems to have paid much attention to it and, as I have argued elsewhere, it does not seem to have been directed at Native fishers.[41] For those who might have been concerned about a lack of protection for Native fisheries, the fisheries clause in the Douglas Treaties and a somewhat weaker, although nonetheless prevalent, sense of Native fishing rights would probably have seemed enough. The experience of Native peoples around the Great Lakes suggested otherwise.

Fisheries Law and Policy in Pre-Confederation Canada

The colonies of Vancouver Island and British Columbia had relatively little contact with the Province of Canada until the negotiations that would lead British Columbia into the Canadian confederation in 1871. Colonial officials had much more to do with HBC headquarters and the Colonial Office, both of which were based in London, than they did with the eastern colonies. Indeed, it was through the Colonial Office that Douglas received copies of the fisheries legislation from the Province of Canada in the early 1860s.[42] As a result of these tenuous connections, Indian land policy and fisheries regulation in the east had little impact in British Columbia until after Confederation. However, when British Columbia joined Canada, ceding jurisdiction over "Seacoast and Inland Fisheries" and "Indians and Lands Reserved for Indians" to the Dominion government, it received Dominion fisheries law as well as the existing bureaucracies for managing fish and the lives of Native peoples.[43]

The province's refusal to recognize Native title, and its insistence that its land policy continue after British Columbia joined Canada in 1871, provoked considerable debate between the Dominion and the province. The introduction of Dominion fisheries law and a bureaucracy to manage the fishery proved much less contentious. In the mid-1870s, the operators of the industrial commercial fishery at the mouth of the Fraser River began clamouring for state regulation of the salmon fishery, at least to limit competition. As a result, the Dominion introduced the *Fisheries Act* of 1868 to British Columbia in 1877[44] and the first set of regulations for the province in 1878.[45] Fisheries did little in the 1870s to enforce these laws, and Native fishers were informally exempted from the regulations, but the beginnings of the legal framework that would govern the fishery were in place. This legal framework, and the bureaucracy that the Dominion established to enforce it, had emerged out of a history of conflict with Native fishers in the Great Lakes. Many of the important figures in the Fisheries Branch of the Department of Marine and Fisheries, including the commissioner of Fisheries, W.F. Whitcher, had cut their teeth on the conflicts that ensued when long-established Native fisheries, most of them protected under treaty, were reallocated to non-Native commercial and sport fishers. It was this legacy of conflict that British Columbia acquired when the Dominion assumed responsibility for the fisheries.

Crown policy on the Great Lakes had not always been thus. In the early nineteenth century, as the non-Native population spread north and west from Lake Ontario, the British signed a series of land cession treaties with the Ojibwa. Historian Michael Thoms, drawing on oral histories and historical records, argues that these treaties were intended to allow for the political, economic, and ecological coexistence of the Ojibwa and the newcomers. The Ojibwa reserved the wetlands, sites of their most productive fisheries, and ceded the arable uplands to the newcomers to support their agrarian-based economy. Although the written texts of most of these early treaties do not mention fish, the Legislative Assembly of Upper Canada followed every treaty in the early nineteenth century with legislation prohibiting non-Native access to the productive fisheries at the mouths of the rivers flowing into Lake Ontario. These fisheries acts, Thoms suggests, are evidence of the colonial state fulfilling its obligations, under the treaties, to protect the Ojibwa's exclusive fisheries by prohibiting non-Native access at the times and places where the Ojibwa fished. He finds the antecedents of this legislation in English fisheries acts that set aside the inland fishery for the landed gentry. The Upper Canadian legislation of the early nineteenth century, he concludes, was similarly intended to preserve the fisheries for a particular group – the Ojibwa.[46]

As non-Native interest in the fisheries grew, however, so did incursions into Ojibwa fishing grounds. Despite well-documented Native protests, growing more vigorous in the 1830s, that the existing laws were inadequate and insufficiently enforced to protect the Ojibwa fisheries, government officials did little to exclude non-Native fishers. In fact, the colonial government began issuing commercial fishing licences to non-Native fishers in waters that the Ojibwa had thought were reserved, under treaty, exclusively to them. In some cases the government required the licensee to secure permission from and compensate the Ojibwa for the use of their fishing grounds, but these stipulations were seldom enforced. Historian Victor Lytwyn has chronicled the growing conflict over fisheries around Manitoulin Island and the Saugeen Peninsula on Lake Huron in the mid-nineteenth century. This conflict between Ojibwa fishers, non-Native licence holders, and government officials would include the alleged murder of Fisheries Overseer William Gibbard in 1863.[47] Legal scholar Peggy Blair finds the roots of this conflict in a fundamental transition in government policy in the mid-nineteenth century, perhaps best evidenced by the widely circulated 1845 opinion of Attorney General W.H. Draper: "I have the honor to report my opinion, that the right to fish in public navigable waters in Her Majesty's dominions is a common public right – not a regal franchise – and I do not understand any claim the Indians can have to its exclusive enjoyment."[48] In this view, rights to exclusive fisheries derived from the parliament; Native rights to exclusive fisheries that preceded the Crown's assertion of sovereignty, even if recognized by the Crown in treaties, were unenforceable at law unless approved by parliament.[49] The accuracy of this opinion has been strongly challenged in recent work,[50] but even if it were an accurate summation of the law, Thoms argues that parliament gave its imprimatur to the exclusive Ojibwa fisheries in the legislation that followed the treaties.[51]

Government officials had either forgotten or dismissed the earlier treaty promises when, in 1857, the Legislative Assembly for the Province of Canada passed its first comprehensive fishing legislation, at the same time rescinding the earlier acts that had been intended, at least initially, to protect Ojibwa fishing.[52] The new *Fishery Act*, substantially revised (and renamed *Fisheries Act*) in 1858, offered commercial fishers greater security of tenure to fishing grounds through Crown-issued licences and leases.[53] It also set aside particular fish for the benefit of sport fishers, who, Thoms suggests, were the most important lobby group behind the new legislation. The new acts did not mention or protect the Ojibwa treaty rights to fish. Indeed, Ojibwa fisheries in Lake Huron were soon overwhelmed by non-Native commercial operators brandishing Crown-granted licences and leases for shore-based seine-net fisheries. At the behest of sport fishers, Fisheries officers targeted Ojibwa fishing in the smaller lakes and rivers as well, imposing close seasons and gear restrictions that accommodated sport fishers' interests and, in effect, setting aside the fisheries for them. In 1865, the government added a section to the *Fisheries Act* that allowed officials to permit Indian food fishing with otherwise prohibited technology at times and in locations that were closed to other fishers.[54] Exercised at the discretion of the local Fisheries officers, these food fishing privileges allowed restricted opportunities for Native fishers to operate in territories that they had never ceded under treaty, or in which they thought they had treaty rights to exclusive fisheries. In short, it was this institutional history of conflict over fish, and the substantial marginalization of the Native fishery, that British Columbia inherited when the Department of Fisheries assumed control of the Pacific coast fisheries in the mid 1870s.

Conclusion

When, in 1871, British Columbia joined the Canadian confederation, the negotiated terms of union gave the Dominion government jurisdiction over both fisheries and Indians, placing responsibility for them in the departments of Fisheries and Indian Affairs. The terms of union also required the Dominion to pursue as liberal a policy towards Indians as had the former colony.[55] The Conservative government of John A. Macdonald was soon to realize, however, that the colonial policy in British Columbia had become neither liberal nor generous, and that the province refused to recognize Native title or the need to extinguish it through treaty. The colonial land policy, as it had evolved under Joseph Trutch, of small, scattered reserves that provided some minimal protection for Native villages and resource procurement sites, became the land policy in the province of British Columbia. This policy would be tempered somewhat by a Dominion government that sought some additional accommodations but that would not press the issue of Native title in negotiations with the province or through the courts. Reserves remained small.

Although the provincial land policy prevailed, it was the Dominion's legislation, regulations, and policies that governed the fisheries. Built upon the common inheritance of the English common law, a Canadian fisheries regime, which had been overtaken in the mid-nineteenth century by a revisionist understanding of Native treaty rights to fish in the Great Lakes, arrived in British Columbia. Specific regulations tailored for British Columbia would soon appear, and local fisheries officials would inflect the law towards local circumstances through its enforcement or, initially, lack of enforcement. Nonetheless, the essential legal elements and an institutional history that had been formed by a culture of conflict with Native fishers, and their dispossession, were now in place in British Columbia.

All but forgotten, or at least ignored by officials within the Department of Fisheries, was the fisheries provision in the Douglas Treaties. In securing the right to "fisheries as formerly," the treaties recognized the fundamental importance of the fisheries to the lives, economies, and cultures of Native peoples on the Pacific coast. Under the treaties, the fisheries were, as Douglas thought they should be, "fully secured to them by law." Native peoples were soon to discover that this was not the case, or that, if it were, legal protection was inadequate to secure to them their fisheries.

2
Land Follows Fish

The fisheries had not been sites of conflict between Native peoples and newcomers in pre-Confederation British Columbia. Native fishers continued much as they had done, taking advantage of new markets for their fish, but without interference from the colonial state or competition from non-Native settlers. That changed dramatically on the Fraser River in the 1870s with the arrival of canning technology and the rise of the industrial commercial fishery. The first salmon cannery appeared at New Westminster, near the mouth of the Fraser, in 1871. Three canneries were operating there in 1876, two more the next year, and eight by 1878. Several canneries had also opened in the north near the mouth of the Skeena River, and there were salting operations along the coast for salmon and other species.[1] The Hudson's Bay Company had shipped barrels of salted salmon around the Pacific, but canned fish could reach the markets in eastern North America and Europe, and production increased rapidly in the last quarter of the nineteenth century, putting new pressures on the resource.[2] Whether Native peoples participated in the industrial commercial fishery or not, the new concentration of capital changed their fisheries. For some, it provided new opportunities and increased incomes; for others, it reduced access and created considerable hardship.[3] In either event, management of the fishery moved beyond Native control to the Dominion Department of Marine and Fisheries (Fisheries). As a result, the fisheries became, and remain, one of the principal sites of conflict between Native peoples and the government of Canada.

The 1870s brought other important changes as well, particularly a growing non-Native population and its increasing occupation of Native peoples' land. In 1871 there were approximately 36,000 people living in the province, roughly 30 percent (10,500) of whom were immigrants.[4] Most of these new arrivals lived in or around the townsites of Victoria and New Westminster. Ten years later the population had increased to 50,000, nearly half of them immigrants. Although the non-Native population remained concentrated on southern Vancouver Island and in the Fraser Valley, it had begun to spread inland along the Fraser, Thompson, and Nicola rivers and into the Okanagan.[5] The Crown distributed land to these settlers, either through sale or pre-emption (a process to acquire land before its survey), and although the total acreage of alienated land was small, there was not much arable land in the province.[6] In the absence of treaties or any other arrangement with Native peoples over land, many of these grants produced conflict. Occasionally the conflict erupted into violent confrontation, the most notorious of which was the killing of eighteen non-Native road builders at Bute Inlet in 1862 and the subsequent hanging of five Tsilhqot'in.[7] Events such as this, as well as the "Indian Wars" in the Washington Territory or the earlier shelling of Native villages by Royal Navy gunboats, were the dramatic events that lay in the background of the pervasive violence that marked the enclosure of traditional Native territories. On the ground, fences followed cadastral surveys and maps, and the property lines they created were backed by state law, in turn buttressed by violence. "The establishment of a Western liberal property regime," writes Nick Blomley, "was both the point of these violences and the means by which violent forms of regulation were enacted and reproduced."[8]

Among these lines of private property imposed on the land and the people who lived on it, the Dominion and provincial governments also created boundaries that marked the spaces that were to be reserved for Native peoples. This chapter examines the crucial role of the fisheries in the first attempts to resolve what had become known, following the publication in 1875 of a book of letters and other documents under the same title, as the "Indian Land Question."[9] That question would be answered in British Columbia with small, scattered Indian reserves premised on access to fish.

The Joint Indian Reserve Commission, 1876-78

The governments of Canada and British Columbia had not been able to agree on many issues after Confederation, including the size of Indian reserves and the underlying issue of Native title. As a result, nothing happened on the Indian land question between 1871 and 1875 except the alienation of more land to non-Native settlers. Finally, in 1876, amidst growing unrest among Native peoples over the occupation of their land, the two governments established a commission to investigate the Indian land question. The Joint Indian Reserve Commission (JIRC) was the first attempt by the Dominion and province to seek a satisfactory resolution. Each government would appoint a commissioner, a third would be appointed jointly, and together the commissioners would travel the province to gather evidence about the Native population, consider existing land use, and make recommendations about what parcels of unalienated Crown land should be set aside as Indian reserves. Land that had been sold or that was subject to a pre-emption claim would not be considered. The question of Native title, which the province refused to acknowledge and the Dominion was prepared to overlook, was set aside.[10]

After considerable delay, the province appointed Archibald McKinley, a retired Hudson's Bay Company trader, as its representative and agreed with the Dominion to appoint Gilbert Malcolm Sproat as the joint commissioner. The Dominion had earlier appointed Alexander Caulfield Anderson, also a former Hudson's Bay Company employee and the first inspector of Fisheries for the province of British Columbia, his dual appointment emphasizing the perception within the Dominion government that the allotment of Indian reserves was closely tied to the fisheries.

The commissioners began their work late in 1876, operating on instructions from the two levels of government about the terms of their commission and the role of the JIRC, including its treatment of Native fisheries. David Laird, the Dominion minister responsible for Indian Affairs, told Anderson that although the long-term goal was to create self-contained agrarian communities, he was not to disturb those Indians engaged in "any profitable branch of industry," including fishing:

> While it appears theoretically desirable as a matter of general policy to diminish the number of small reserves held by any Indian nation, and when circumstances will permit to concentrate them on three or four large reserves, thus making them more accessible to missionaries and school teachers, you should be careful not even for this purpose to do any needless violence to existing tribal arrangements, and especially not to disturb the Indians in the possession of any villages, *fishing stations,* fur-trading posts, settlements or clearings, which they may occupy and to which they may be specially attached, and which may be to their interest to retain. Again it would not be politic to attempt to make any violent or sudden change in the habits of the Indians, or that those who are now engaged in *fishing,* stock-raising, or in any other profitable branch of industry should be diverted from their present occupation or pursuits, and in order to induce them to turn their attention to agriculture.[11]

Two years earlier, Dominion officials had approved a comparable position by order-in-council: "Great care should be taken that the Indians, especially those inhabiting the Coast, should not be disturbed in the enjoyment of their customary *fishing grounds,* which should be reserved for them previous to white settlement in the immediate vicinity of such localities."[12]

The provincial government adopted a similar approach. In an 1875 report on Indian reserves, the Attorney General, G.A. Walkem, noted that most Native people in the province lived in fishing communities and therefore needed access to their fisheries rather than agricultural land. He wrote that "it is reasonable to suppose that large tracts of agricultural land will not be required for the class of Indians referred to. Those who cannot be employed usefully ... in *fishing* or hunting, might require and fairly expect farming lands. The other portion of the community would be provided for in other ways, by reserving their *fishing stations,* fur-trading posts and settlements, and by laying off a liberal quantity of land for a future town-site."[13] As a result, the province instructed its appointee, McKinley, and the joint appointee, Sproat, not to disturb Indian fisheries: "You will avoid disturbing them in any of their proper and legitimate avocations whether of the chase or of *fishing,* whether pastoral or agricultural."[14]

These instructions to the reserve commissioners contained varying terminology. Sometimes a "fishing ground" was to be protected, other times a "fishing station," and sometimes the activity itself – fishing. The reserve commissioners generally described those allotments made primarily to protect fisheries as "fishing stations," a term that the Supreme Court of Canada has recently interpreted to refer to points of land, but not to the adjacent waters where fish were caught.[15] Although plausible, such a clear distinction between land and fisheries is problematic. Many important indigenous fishing technologies, such as weirs, dip nets, and reef nets, were land-based and, as such, were inseparable from the adjoining land, foreshore, or bed of the body of water. If a fishing station were allotted to secure a weir fishery, as many were, then it is reasonable to suppose the reserve commissioners intended that the reserve include that portion of the river in which the fence-like structure was built, or, at very least, that the reserve include protection for the weir fishery.[16]

The uses of the term "fishing station" vary in the nineteenth century. Sometimes it seems to refer only to the dry land from which fisheries were conducted; at other times it refers to both the land and water, or only to the water on which people fished. As an example of the latter, John McCuaig, the superintendent of Fisheries for Upper Canada, wrote in his annual reports for 1858 and 1859 that he hoped to lease fishing stations in waters adjacent to lands that were held by someone else. This he found difficult: "Another obstacle to the profitable leasing of the Fisheries is found in the refusal of some of the parties owning land on the waters edge to allow a landing place to the fishermen, who might otherwise be willing to lease *stations* in front of such properties."[17] In short, fishing stations might include or exist independently of dry land.[18] When, in 1901, British Columbia passed its first *Fisheries Act,* the word "station" was used in a manner that suggested both land and the adjacent fishery: "'Fishery' shall mean and include the particular locality, place or station in or on which a seine, pound or other net is used, placed or located, and the particular stretch of waters in or from which fish may be taken."[19]

A practical reality of the fishery in the nineteenth century, an era of human- and wind-powered fish boats, was that the control of particular points of land effectively secured control of many fisheries. With some exceptions, such as the mouth of the Fraser River, where current, tides, and waves were moderate enough to allow some mobility, access to identified outcroppings or landings was essential to a successful fishery in British Columbia and elsewhere.[20] This was particularly true on many rivers, but also on long stretches of a rough and treacherous west coast. It was only in the twentieth century, with the proliferation of gasoline-powered boats and the increasing enforcement of laws that restricted land-based fishing technologies, that control of particular points of land became less important.[21]

Terminology aside, the provincial government's goal was to reserve as little land as was necessary to quell Native unrest and at the same time ease Native peoples into the wage economy. Officials hoped that the reserves would be temporary, a bridge that would help Native peoples make the transition from traditional to modern economies. They were not to be consolidated on large reserves, but were to have some limited space in their traditional territories from which to access their traditional food sources while gradually integrating, with the help of missionaries, into the wage labour force, particularly that of the emerging industrial commercial fishery. This was the provincial model to which Dominion officials were willing to accede. The Department of Indian Affairs emphasized that the reserve communities were to be self-sustaining, but this reflected as much a desire to keep Native peoples off their welfare rolls as it did a fundamentally different vision. As a result, the land policy was built around access to the fisheries. The governments could allot small parcels of land, leaving most of the province open for non-Native settlement and development, with a reasonable expectation that the fisheries would remain central to Native economies as a means of subsistence but also as a site of their wage labour.

The commissioners quickly discovered, however, that although the canneries offered new employment opportunities, the rapidly expanding industry also threatened many Native fisheries. Cannery fish boats, many of which were worked in the early years by Aboriginal fishers, occupied important fishing grounds, their numbers and nets precluding those who claimed prior rights of access. The Indian superintendent for British Columbia, I.W. Powell, called for Fisheries to curb the aggressive cannery-based fisheries that were producing conflict with Natives, and for the JIRC to set aside Native fishing sites. Indian Affairs, he argued, "should take steps as soon as possible to reserve certain *fishing grounds* for the Indians who will be sure to create trouble if not thus cared for."[22]

Inclement weather prevented long excursions in the winter of 1876-77, but the commissioners did travel from New Westminster to Musqueam at the mouth of the Fraser, around Burrard Inlet, and then up Howe Sound and Jervis Inlet, across the Strait of Georgia to Vancouver Island, and south along the island's eastern shore. As they went, the commissioners established procedures for allocating reserves and recording their decisions. The "Minutes of Decision" included a description of the reserve boundaries and a sketch, intended to guide the surveyors who would follow. In some cases they gave the reason for granting the reserve. The more general descriptions of the reserves and their uses, often written some months later, became known as "Field Minutes."

The commissioners specifically linked many of the reserves granted on their first circuit to Native fisheries. In the Cowichan Valley, for example, they described four reserves along the upper Cowichan River, at weir-fishing sites, as fishing stations. These explicit connections, however, do not convey a complete accounting of the ties between fisheries and reserves. The large Cowichan Indian Reserve No. 1 at the mouth of the river, which included the principal winter village sites of six communities and their fish weirs on the Cowichan and Koksilah rivers, was also intended to protect those fisheries, although the commissioners did not formally record the importance of these fisheries or the correlation between them and this reserve.

Shortly after the JIRC's visit, the province granted a portion of the large Cowichan reserve to William Sutton, who proposed building a lumber mill and running logs down the Cowichan River to it. Sproat vociferously denounced the province and its disregard of the commission's work. However, if the land grant were to stand without the Cowichan's consent (Sproat thought it should not), then the Cowichan, he argued, must be compensated not only for the land, but also for the loss of their fishery, which would be destroyed by the log runs. According to Sproat, he and the other commissioners on the JIRC had promised the Cowichan that their fishery would not be interfered with or disturbed. The loss, therefore, of such an important weir fishery required Cowichan consent and, if necessary, compensation. The purchaser, who had already paid the Crown, agreed to purchase it again from the Cowichan and with it, perhaps, the right to run logs down the river, although that was not specified in the purchase agreement. What is clear, however, is that the JIRC had made specific representations to the Cowichan and to the other bands it consulted on this initial circuit that their fisheries would not be disturbed. Disruptions to Native fisheries, such as the Cowichan weir fishery, required consent and compensation. The Cowichan certainly believed that their fisheries had been protected, and later in the nineteenth century, when officials challenged the Cowichan right to build fish weirs, the Cowichan repeatedly referred to the JIRC guarantees as one of the sources of their continuing right.[23]

In 1877, the commissioners travelled on their second circuit from Kamloops Lake along the Thompson River to Shuswap Lake, and then south through the Okanagan to the border with the United States. The influx of non-Native settlers into this region before negotiations and treaties over land had produced considerable tension, and the possibility of an "Indian War" was in the air. Land and water were issues everywhere, but so was access to salmon and trout fisheries, and the commissioners were divided on what they might do to protect indigenous fisheries.[24] The JIRC had no mandate to disturb settlers from land that they had acquired by purchase or pre-emption. It was illegal, however, for settlers to occupy Native settlements.[25] The commissioners were prepared to include such illegally alienated land in reserve grants, and they did so, after much discussion, at the north end of Okanagan Lake.[26] Anderson, the Dominion commissioner, argued that Native fishing sites were settlements, that settler occupation of these settlements was illegal, and that the land should be returned to its Native occupants whether or not it had been pre-empted or purchased from the provincial government. He wrote to Fisheries: "In all cases, so far, their [Indian] interests as fishermen have received my best attention; and I shall continue to urge upon my brother Commissioners whose views on this subject, I may add, do not differ from my own, the necessity of securing the hereditary privileges of the Indians wherever they may appear to be imperilled."[27] Anderson may have overstated the consensus on the commission. McKinley, offering a provincial voice, argued that indigenous fishing sites were too numerous and would unduly restrict white settlement if they were all protected within reserves. Settler access to water for irrigation was also important, and this might be compromised if all the many seasonally important trout fisheries in the small lakes and rivers of the interior were set aside as reserves.

The commissioners, it seems, could not agree whether Native fisheries constituted settlements that limited non-Native settlement. They sought new instructions. In the meantime, Sproat proposed a way out. The commissioners' first priority, following their instructions, he argued, was to protect Native fisheries, but they also had to recognize that an incoming settler society wanted land for farming or ranching, and water for irrigation. Where interests conflicted, Sproat suggested that "if the Indians were in good humour and gave consent, and if it were clear that they were merely resorting to some of these places from habit and not to supply essential needs, the Dominion Government, possibly might consent to the Provincial Government buying out the fishing rights of the Indians in places where it was desirable in the public interest to release water for irrigation."[28] In other words, the province might negotiate to purchase Native fishing rights, with the consent of both the band and the Dominion required for an agreement; otherwise, in Sproat's view, Native fisheries were secure. In effect, Sproat was proposing a collection of tripartite fish and water treaties between Indian bands, the Dominion, and the province if settlers were to have access to these important resources. It was one of Sproat's many well-intentioned attempts to find common ground.

Native peoples in the interior clearly believed that the JIRC had promised them continuing access to and control of their fisheries. In 1906, Fisheries Guardian John T. Edwards reported on his efforts to remove fish traps and weirs along the North Thompson. The Indians, he reported, were determined to maintain their traps at the mouth of the river, at Kamloops, and refused all entreaties to dismantle them. Not only did the traps secure a necessary food supply, argued an unidentified North Thompson chief, but the river belonged to them. "I don't think that you or anyone else," he told Edwards, "has any right to interfere with our mode of fishing on that stream. That stream ... belongs to us. The Queen gave us that stream years ago." Edwards wrote: "I tried to reason with him, that the good old Queen gave them five acres of ground at the mouth of the barriers to fish upon, but not to put any traps on the same stream, but he wouldn't have it that way."[29] It is likely that the chief was referring to his people's understanding of the representations made to them by the JIRC, or perhaps to earlier representations made by Governor James Douglas. In any event, this understanding, widely held, coincides with what the Indian reserve commissioners themselves said they were doing.[30]

However, the JIRC was not to continue. The provincial government thought the process too expensive and the commissioners too generous in their reserve allotments. The JIRC was disbanded, but the provincial government reluctantly agreed, after it became apparent that the diplomacy of the JIRC had probably averted a serious uprising in the interior, that a reduced commission would continue the work. It was prepared to allow the joint commissioner, Sproat, to act as the sole commissioner. Before he returned to the field, Sproat insisted that he have clear authority to make decisions in the field, an authority that the province, after some delay and reluctance, conceded.[31]

Indian Reserves and Fisheries Regulations, 1878

In May 1878, Sproat, now the lone commissioner of the reduced Indian Reserve Commission, began the difficult work of confirming old reserves and allocating new ones in Nlha7kapmx territory (Photo 2.1). He travelled carefully and slowly along the Fraser, Thompson, Nicola, Coldwater, and Similkameen rivers (Figure 2.1), attempting to make the difficult compromises necessary to secure Native space, but working within the constrained parameters which required that he not disturb Crown grants or pre-emption claims or eliminate the possibility of further non-Native settlement. Here, as elsewhere in the interior, access to land and rights to water were the principal points of conflict. Water was needed to irrigate fields, but also for the placer mining operations that dotted the Fraser. Agricultural prospects were marginal, at best. There were precious few locations where flat land combined with access to water to create the possibility of viable farms.[32]

Photo 2.1 Indian Reserve Commissioner Gilbert Malcolm Sproat in front of his tent at Yale, likely in August 1879. | *BC Archives A-01771*

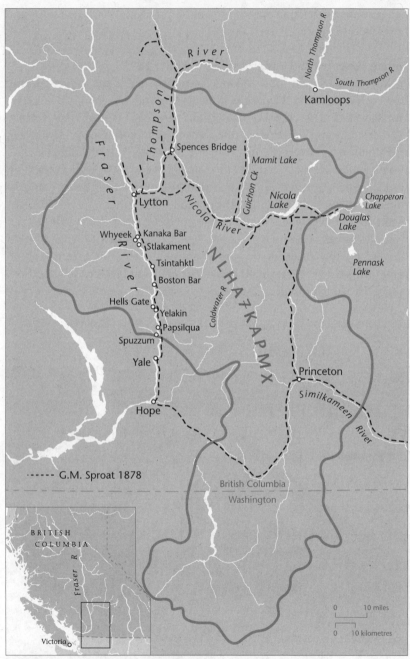

Figure 2.1   Sproat's route through Nlha7kapmx territory, 1878. He began at Hope and headed north, went up the Fraser Canyon to the Thompson and then Nicola rivers, and then south to the Similkameen River. | *Source:* Cole Harris, *Making Native Space* (Vancouver: UBC Press, 2002), 139.

In comparison to the competition for water and land, the fisheries seemed relatively straightforward. Settlers were generally not interested in fishing in the Fraser Canyon, and although the canning industry was expanding rapidly on the coast, there was little direct or local competition for the Nlha7kapmx fisheries. Of course, the claims to land, water, and fish overlapped. If water were diverted for irrigation or placer mining, it reduced stream flows and damaged or eliminated trout fisheries in the tributaries of the Fraser. Access to salmon fisheries along the Fraser itself was also an important and potentially vexing issue. The canyon did not provide uniform fishing opportunities. The Nlha7kapmx fished at certain times of year in particular locations where their technology, primarily the dip net, was most effective. Working a back eddy where salmon rested out of the main current, a dip-net fisher could catch several hundred fish a day (Photo 2.2). The salmon would then be dried on a rack built on a crag above the fishing site.[33] If a settler pre-empted or purchased land that included a fishing site, Nlha7kapmx fishers might find the land fenced and their way barred. If they broke the fences, they were liable for damages for trespass, including

Photo 2.2   Native fisher working a back eddy with a dip net from a wooden platform in the Fraser Canyon, ca. 1890. Note the head and pole of another fisher working just upstream, the dip net lying across the scaffolding in the foreground, and the drying racks above and across the river. | *BC Archives A-08917*

Photo 2.3   Indian Reserve Commission camp at Spuzzum, May 1878. Commissioner Sproat stands with the flagpole on his right and commission census taker George Blenkinsop to his left. Standing and seated before them are members of the Spuzzum band. | *BC Archives A-10772*

fence repairs, damage to crops, and the costs of escaped animals. The fisheries themselves, however, were relatively uncontested.

At Spuzzum, Sproat confirmed the four existing reserves and added several others, including two additions to the reserve at Yelakin (Photo 2.3). The first addition began on the Fraser at the northwest corner of the old reserve and continued up the river more than half a mile, but it was to be a narrow strip – "of such a width and no more, as considering the height of the waterline in the fishing seasons shall in the surveyors judgment enable the Indians to carry on their fisheries as heretofore and to have access to said strip from the present reserve."[34] A similarly narrow although somewhat longer strip was to extend south from the reserve along the east bank of the Fraser. Including the old reserve, Sproat had set aside a little over two miles of the Fraser's east bank for the Spuzzum fishery. Farther south, between the Fraser and the Papsilqua reserve, which occupied a small bench above the river, Sproat allotted another five-acre fishing station. These fishing reserves were intended to set aside the principal fisheries; Sproat did not intend that the Spuzzum fishery would be restricted to these places, but only that they would have the additional protection of being reserves. None of these fishing reserves were confirmed, but almost forty years later the Royal Commission on Indian Affairs for the Province of British Columbia allotted reserves on either side of the Papsilqua reserve, bordering on the Fraser (see Appendix, Figure A2).

At Boston Bar, Sproat confirmed two existing reserves and recommended a number of others, including several small fishing stations and the major fishery at Hells Gate (Figure 2.2). The reserve at Hells Gate was to be small – five acres on the west bank and ten acres on the east – but there were complicating factors. Sproat noted a small house on the east side of the river on the proposed reserve. It was occupied by John Mancell, who claimed that he had lived in it for twelve years as the section man on the Cariboo wagon road and that he had cleared, fenced, and otherwise improved the land around the house. The Nlha7kapmx, he asserted, used the site as a summer fishery, and he had not interfered with their accustomed use. Sproat responded that it was clear Mancell had a moral claim, although not a legal claim, to the house and surrounding garden. The land had been pre-empted and a certificate of improvement issued to another party in 1865, but no Crown grant had been issued and that party was no longer around. Even if the original claimant had been interested, or had formally transferred his interest to Mancell, the pre-emption claim was of no force, Sproat suggested, to the extent that it included "the old Indian fishery and what naturally and reasonably belongs to it." In other words, Native settlements, including important fisheries, could not be legally pre-empted or purchased. Sproat indicated that he did not propose to mark off all Indian fisheries along the Fraser as reserves – only those principal sites of particular importance. The fishery at Hells Gate was one such site. Sproat hoped, nonetheless, that he could meet with Mancell and that a compromise might be reached.[35] It is not clear from the records whether that meeting ever took place, but the reserve was never confirmed. Thirty-six years later, a slide at Hells Gate, precipitated by blasting done for the construction of the Canadian Northern Railway, devastated the Fraser sockeye runs.[36]

Sproat had expressed some concern that allocating reserves as he was doing might not be the best way to secure Native fisheries. He wrote that the reserves he had allotted at Boston Bar "are laid off subject to the result of consultation by the Commissioner with the governments, as to whether this is the only way of securing the principal Salmon fisheries of these Indians."[37] Without opportunity for such consultation, he continued in a similar vein. For the Boothroyd band, he extended the Tsintahktl reserve, initially allotted by the Royal Engineers in 1861, to include a fishery. He

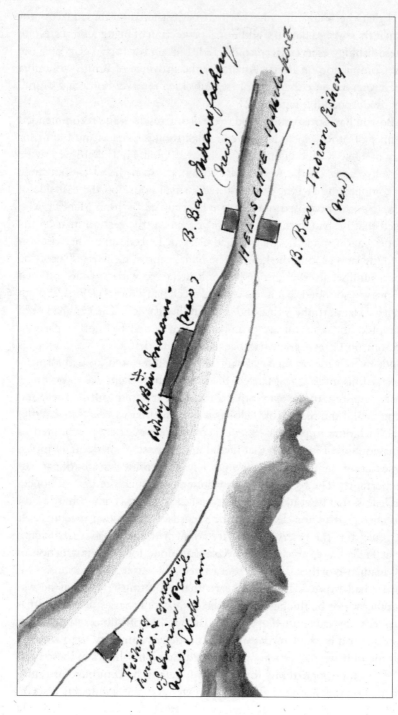

Figure 2.2 Sproat's sketch of proposed reserves at Hells Gate, 1878. | *Source:* Minutes of Decision, 1 June 1878, Federal Collection of Minutes of Decision, Correspondence and Sketches (Fed. Col.), vol. 6 (G.M. Sproat, May 1878 to Aug. 1878), p.32.

also recommended a five-acre reserve to border on the Fraser below the Stlakament reserve, which occupied a small bench above the river. This fishing reserve was never confirmed.

At Kanaka Bar, Sproat encountered another land-use conflict involving Nlha7kapmx fisheries. First, he confirmed and added to the reserve allotted by the Royal Engineers in 1861. He also granted the 350-acre Whyeek reserve on the west bank. At the south end of the Whyeek reserve, or perhaps just south of the boundary – Sproat could not be sure until the land was surveyed – he reserved an important fishery. Directly across the river, on the east bank, was another significant fishery, but this land was owned by an Italian settler named Palmer (also referred to as Parma or Palma). Sproat tried to insert an easement against Palmer's title to ensure Nlha7kapmx access (Figure 2.3):

> The Indians are to have a right of fishing along the whole of Palma's frontage and the surveyor will arrange for suitable access and mark same on ground and on plan, as most convenient for Palma, so as not to cross cultivated land, or unnecessarily spoil fences. The Indians may be reminded that I only gave them fishing and access here.[38]

This was another example of Sproat's attempt at compromise. But as with the others, there is no indication that this compromise was ever formalized. The easement does not appear to have been recorded.

Several days earlier, Sproat had attempted to broker a similar compromise, albeit from a greater distance. The JIRC had set aside a large reserve for the Kamloops band, part of the Secwepemc tribal group, at the junction of the South and North Thompson rivers. It had also set aside several small fishing stations. One of these stations – Kamloops Indian Reserve No. 2 – was intended to secure an important weir fishery on a creek south of Kamloops, but the fishery was encompassed by land that had been pre-empted by Bartlett Newman in 1873. The fishery was particularly important and must be protected, Sproat argued, not only because it was a Native settlement and therefore not eligible for pre-emption, but also because he had persuaded the Kamloops to abandon another fishery (he did not indicate which one) if that within Newman's pre-emption was protected. However, Sproat thought that the reserved fishery would interfere with good hay land, which he was reluctant to do, and that the Kamloops might be induced to relinquish their claim to it if they were compensated. He believed the sum of fifty dollars might be sufficient. But the Kamloops were not interested in alienating their fishery, so Sproat proposed another

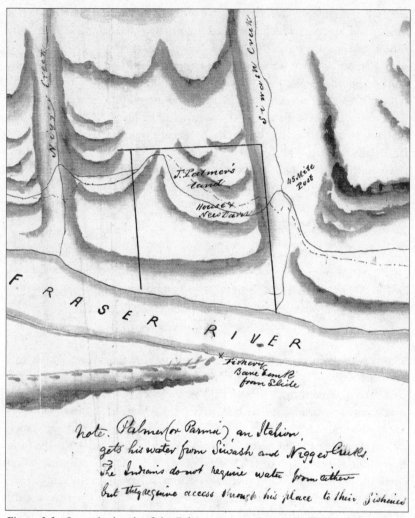

Figure 2.3 Sproat's sketch of the Palmer pre-emption and a Native fishing site on the Fraser River. The notation reveals the importance of water and fish in the process of reserve allotments. "Note: Palmer (or Parma), an Italian, gets his water from Siwash and Nigger Creeks. The Indians do not require water from either but they require access through his place to their fisheries." | *Source:* Minutes of Decision, 1 June 1878, Fed. Col., vol. 6, p. 66.

arrangement in an effort to accommodate farming and fishing interests: exclude the fishery from the reserve, but create a small area near but not at the fishery where the Kamloops could camp and impound their horses (Figure 2.4). That way, the Kamloops could fish and Newman could farm. Sproat offered the following rationale:

Unquestionably, if the matter were pressed the fishery (though it seems a hard thing to say of such a place) would appear to be an old settlement, and any pre-emption of land including it would be invalid. But neither Govt. wishes to raise such a question. The fishery must be laid off with the least inconvenience to all parties.

From this point of view, an enclosure for the horses of the Indians when visiting the fishery is desirable, so that they can have a pretext for trespassing on the lands of the white settlers for grass for their animals. This enclosure generally speaking should be actually at the fishery, but in this case, it might do at a little distance and they should make their camps within the enclosure ...

The Indians may wish for some of the low ground immediately at the fishery, but this Mr Newman requires for cultivation, and I have to give the Indians a fishery there, not arable land, but only an enclosure as above states.[39]

Sproat sought compromise, workable solutions, and "the least inconvenience to all parties," but some of the competing interests were difficult to reconcile.[40] In this case, the province ignored Sproat's efforts to accommodate the Native fishery, just as it did in almost all the other instances where the proposed resolution might diminish a settler's property rights. In 1889, the province issued Newman a Crown grant without any concession to the Native fishery. The Kamloops Indian Reserve No. 2 included a finger of land that surrounded a small creek and extended into Newman's property (Figure 2.5). However, it did not appear to include the important fishery on the larger creek that Sproat had sought to protect (Figure 2.4).[41]

While Sproat was struggling to resolve difficult issues in the Fraser Canyon and elsewhere, new legal developments threatened his work. On 30 May 1878, only days after he had begun his first circuit, the Department of Fisheries enacted the first set of fisheries regulations for British Columbia. They were exceedingly brief, restricted to salmon, and did not include any mention of the Native fisheries.

1. Drifting with salmon nets shall be confined to tidal waters; and no Salmon net of any kind shall be used for Salmon in fresh waters.

2. Drift nets for Salmon shall not be so fished as to obstruct more than one-third of the width of any river.

Fishing for salmon shall be discontinued from eight o'clock A.M. on Saturdays to midnight on Sundays.[42]

It appears to have taken several weeks for news of the regulations to reach Sproat, but when it did, he responded with astonishment and dismay. The ban on net fishing in fresh waters had serious implications for the

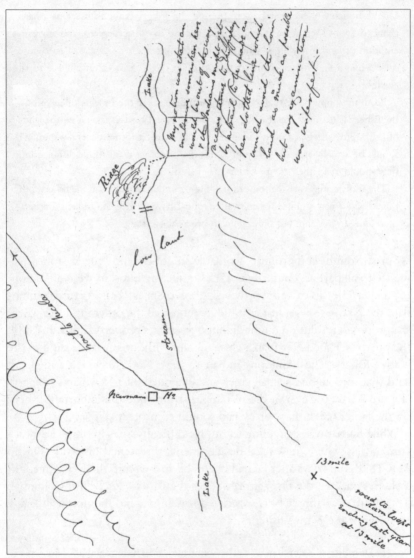

Figure 2.4 Sproat's sketch of Bartlett Newman's pre-emption and the Kamloops fishery, June 1878. Sproat sought to secure the fishery for Kamloops and preserve the good hay land for Newman's ranching operations. He describes the proposed arrangement in the notation at the top: "My notion was that an enclosure some-where here would be out of the way and the Indians might get access thence to the fishery by a trail to be made as per dotted line which would avoid the low land as much as possible but my examination was imperfect." | *Source:* Sproat to E. Mohun, 12 June 1878, Fed. Col., vol. 1 (Letterbook 2).

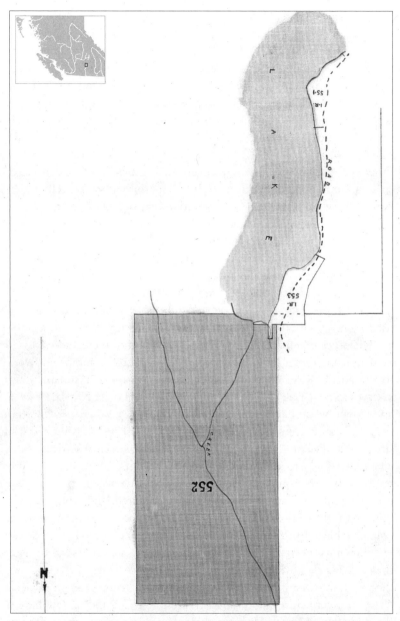

Figure 2.5 Bartlett Newman's Crown grant and the adjacent Kamloops Indian Reserve No. 2 on Trapp Lake. (The map has been inverted to provide the same orientation as Sproat's sketch, Figure 2.4.) Note the finger of the reserve, surrounding the creek, that extends a little way into the Crown grant. However, that finger does not appear to correspond to the location that Sproat identified as the Kamloops fishery on the stream that flows out of Trapp Lake. |
Source: Crown Grant 4091/45, 13 June 1889, Department of Lands, Victoria.

Native fishery. Sproat had just spent more than a month allotting reserves in the Fraser Canyon between Spuzzum and Lytton, where dip nets were the principal fishing technology. Moreover, the records reveal that Sproat had continued where the JIRC had left off, guaranteeing the bands their right to continue fishing undisturbed. "The land question," he wrote, "would be a trifling matter in the eyes of the 30 or 40 thousand Indians who inhabit British Columbia compared with any questions that arose about salmon." Their net fisheries were now prohibited, and Sproat believed that since 30 May 1878 he had "contravened the law" in confirming Native fisheries.[43] He was not the only government official to raise concerns. Inspector of Fisheries Anderson, Indian Superintendent Powell, and even Chief Justice Matthew Begbie expressed dismay over the regulations. Sproat sought new instructions, which he does not appear to have received, but the Department of Fisheries informed Indian Affairs that "the Inspector of Fisheries for British Columbia [Anderson] has been instructed practically to exempt Indians from the operation of the Fishery Regulations of 30th May last, affecting the salmon fishery, in all instances where the fishing is not carried on amongst white people and does not injuriously affect other fishermen."[44] With this informal resolution in hand, Sproat continued much as he had done before, setting aside reserve land to secure access to fish.

At Lytton, Sproat left the Fraser and followed the Thompson River east to the Nicola River where he began working his way upstream, allotting reserves to the Cook's Ferry, Nicomen, Shackan, Lower Nicola, Coldwater, and Upper Nicola bands (see Appendix, Figure A2). Where possible, Sproat secured Native access to their principal fisheries by reserving the fishing sites and sufficient adjoining land to allow for a seasonal settlement. Elsewhere he sought to secure access even if it meant that Natives would have to cross, or fish from, alienated land. To provide access to a tributary of the Nicola River that was surrounded by alienated land, Sproat held that "the old right of the Indians to fish undisturbed on all parts of Mameet [Mamit] Lake and Mameet [Guichon] River and tributaries is declared, but they must not take horses through white man's lands so as to damage them."[45] He then wrote to Joseph Guichon, the owner of a large ranch in the valley, informing him of the easement that encumbered his ownership: "The arrangement is that the Indians are to have, as they always have had, the undisturbed right of fishing in Mameet River and Mameet Lake, with right of access thereto, but in such access they are to walk on foot through the lands of white men and not to go on horseback and they are not to encamp on such lands."[46] Sproat also wrote to two other neighbouring settlers, O'Keefe and Greenhow, that they must "avoid interference with the fishing rights of the Indians as regards fish ascending Meadow Creek [a tributary of Guichon River] in any arrangements you may be making for your mill."[47]

There is no record of the ranchers' responses to Sproat's letters, but it is clear that Sproat's efforts at compromise, his attempts to find space for both indigenous and immigrant inhabitants in the province's interior, concerned many who thought he was conceding too much to the Indians. In 1880, the member of parliament for Yale, Francis Jones Barnard, told parliament that Sproat was being recklessly generous in his reserve allotments. In a speech reproduced in *The Inland Sentinel*, published in Yale, Barnard reported that "the Commissioner, however, seemed to think that all he had to do was to give the Indians whatever land he fancied, whether it was the property of the Government, or of actual *bona fide* settlers; and not only was he to receive this land, but certain fishing rights, or streams, should be secured to him, even if these rights could only be obtained by interference with the whites."[48] Barnard went on to describe how Sproat had denied one settler access to water and another access to fencing timbers in order to reserve fishing stations. There were a great many other concerns raised, all attributed to Sproat's work as the Indian reserve commissioner, but the problems were not of Sproat's making. They were the product of the colony's, and then the province's, earlier failure to address the question of Aboriginal title, or even to develop an Indian land policy, before non-Aboriginal settlement instead of after. Perhaps Barnard was unaware of the compromises, such as those Sproat proposed at Barnett Newman's pre-emption or on the ranches of Guichon, O'Keefe, and Greenhow, where title would remain firmly with the settler, encumbered only by the easement. More likely, these were just the sort of thing that Barnard, as the voice of many, was complaining about.[49]

It seems Sproat intended that Natives should not be disturbed when they prosecuted their fisheries, wherever they were, and whether or not he was able to secure adjacent land as reserve. With deliberate reference to the text of the Douglas Treaties, he indicated that Natives were to carry on their fisheries "as formerly" at the principal fishing sites that he had identified and set aside as reserves, but also more generally. At the end of his field minutes describing the reserve allotments in the Nicola Valley, he wrote:

> The Indians are to have access to, and to be at liberty to carry on, *as formerly,* their fisheries for the various kinds of fish, at their accustomed fishing places, and more particularly in ... [list of locations] ... but the undersigned informed the Indians that with respect to the fish of the Salmon kind, their capture out of season should be discouraged unless required urgently for food, and that the Indians should not at any time destroy salmon roe or take it for use or sale.[50]

The list of locations highlighted particularly important Native fisheries, but Sproat had secured the Nlha7kapmx fisheries, not just those on re-served land.

During his tenure as reserve commissioner, Sproat grew to admire what he understood to have been Governor Douglas's approach to the Native land question.[51] In the late 1870s he sought to implement what Douglas had begun in the 1850s, including the general protection for Native fisher-ies in the Douglas Treaties. The right to "fisheries as formerly," he thought, applied to all Natives in the province, not just the few groups who were parties to those treaties. Sproat suggested that had the government sought treaties with other groups, negotiations over fish, particularly salmon, would have been central. "If the Crown had ever met the Indians of this Province in council with a view to obtain the surrender of their lands for purposes of settlement," he wrote in July 1878, "the Indians would in the first place have made stipulations about their rights to get salmon to supply their particular requirements, and ... land and water for irrigating it would have been, in their mind, secondary considerations."[52] And in November 1878 he wrote: "They have had no treaties made with them, and we are trying to compromise all matters without treaty making. Had treaties been made, stipulations as to salmon would have been in the front. It is, in the absence of treaties, all the more necessary to recognize the actual requirements of the people."[53] Failing treaties, and Sproat was equivocal about whether they would have been the best way to proceed,[54] he developed a system of reserve allotments with access to fisheries at its core. The other elements included a significant land base, access to other resources such as timber, and a measure of self-government that together would enable Native peoples to join the modern, increasingly industrial economy as full participants.[55] But the fisheries, particularly salmon, were central.

Sproat returned to the lower Fraser Canyon in August 1878, at the height of the fishery, to finalize the reserves at Yale, just south of the traditional boundary between Nlha7kapmx and Stó:lō peoples. This fishery, one of the most productive on the Fraser system, drew many thousands of Natives from the Fraser Valley and around the Strait of Georgia every summer, and Sproat discovered great concern that it be protected: "The greatest anxiety was shown by all the Indians as to their salmon fisheries above Yale. Not only are the Salmon caught there used for the sustenance of the tribes of the neighbourhood: they are a commodity in intertribal traffic over a great extent of the country."[56] For this fishery, Sproat provided that "the right of these and other Indians who have resorted to the Yale fisheries from time immemorial to have access to, and to encamp upon the banks of Fraser river for the purpose of carrying on their salmon fisheries in their old way on both sides of the Fraser river for 5 miles up from Yale is confirmed so far as the undersigned has authority in the matter."[57] It was a rough and ready solution, constructed quickly in an effort to protect a valuable Native fishery, but not to preclude settler interest in the land. The attempted compromise was vintage Sproat. There were other important fishing sites in the vicinity, to the north and south along the Fraser, that Sproat could not reserve, but again he tried to confirm rights of access during the fishing season.[58] However, he knew his work at Yale was incomplete, and his language suggests some uncertainty about his authority to make such a grant. His successors on the Indian Reserve Commission would try to sort out the detail of the fishing rights at Yale. In doing so, they would intervene in a dispute involving competing Native interests to particular fishing sites, one that continues to this day.[59]

Conclusion

In 1918, forty years after Sproat's visit, Nlha7kapmx chiefs from the Fraser Canyon wrote to Indian Affairs to protest the increasing enforcement of fish and game regulations and the requirement that their people hold fishing and hunting permits issued by the Dominion or provincial governments. In order to allow the commercial fishery on the coast to continue in the wake of the devastating slides at Hells Gate in 1913 and 1914, Fisheries curtailed the Nlha7kapmx salmon fishery in the canyon. The Nlha7kapmx resisted, referring the department to an agreement made between the late David Spintlum (who held the name of a great chief, Spi'ntlam) and representatives of Queen Victoria that guaranteed their unfettered access to the fisheries.[60] It is not clear from the brief newspaper account of the petition what agreement the chiefs had in mind. Governor James Douglas had been through Nlha7kapmx territory in 1860, explaining in gatherings at Lillooet, Lytton, and in the Similkameen that Indian reserves would be set aside and that the Nlha7kapmx "might freely exercise and enjoy the rights of fishing the Lakes and Rivers."[61] Perhaps this was the agreement that the Nlha7kapmx remembered. If so, the Nlha7kapmx sense that their fisheries had been protected would only have been confirmed by Sproat's visits in 1878 and 1879, and by the later visits of his successor, Peter O'Reilly, in 1881, 1884, and 1886.

Over the four years from the creation of the JIRC in 1876 to Sproat's resignation in 1880, the pattern of reserve allotments, and the importance of fisheries to those allotments, became clear. The provincial and Dominion governments instructed the reserve commissioners to allot reserve land in ways that minimized the disruption to Native peoples. Large, centralized reserves were rejected. In their place, the commissioners continued the practice, established in pre-Confederation British Columbia, of creating many small reserves that were intended to protect Native settlements and resource procurement sites. The commissioners did make efforts to allot arable land, and everywhere they went they looked for places that had not been alienated that Native peoples might farm. This search collided with the reality of British Columbia's mountainous geography, the increasing presence of a settler society that was occupying available farmland, and the orientation of most Native peoples to their fisheries. There was little enough farmland to begin with, the province had granted much of it to non-Native settlers whom the commissioners were unable or unwilling to displace, and the commissioners sought to ensure space for those settlers who were still arriving. As a result, the reserves were allotted to secure to Native peoples their fisheries.

Sproat's intention was clear – he sought to protect Native fisheries as best he could – and he clearly informed the Nlha7kapmx and other groups that they were to have their fisheries as formerly. But one senses Sproat's lingering uncertainty about his authority to reserve fisheries, or even about the best way to proceed to achieve that goal. In 1880, after he had resigned as reserve commissioner, but while he was finishing some uncompleted files for the Fraser Canyon at Spuzzum, Sproat wrote the following field minutes in a tone that suggested advice to a successor:

> The question of what to do in regard to the fishing places of the Indians along the Fraser, and Thompson is a little difficult.
> They fish in many places but, especially, at certain places, and each family has its rock or station.
> The phrase "nobody will interfere with them" will perhaps not be satisfactory to the Dominion Govt., and it may be necessary to reserve some, at least, of the principal places.
> Some other way of arranging the matter may be found, and this may remain for discussion between the Commissioner and the authorities until the line of the railway is defined; meantime, if declaring these localities to be reserves is the rule to be followed, there are very important fishing places near Rombrôt's which might be secured.[62]

Given the divergence of opinion over Indian reserve allotments between Sproat and the provincial government and even Superintendent of Indian Affairs Powell, Sproat's departure from his position as reserve commissioner appears to have been inevitable. The positions taken on the Indian land question had evolved, but in different directions. While Sproat was increasingly sympathetic to Native voices, provincial officials were increasingly determined to remove or prevent any obstacles to the development of the province, and the Dominion government was not prepared to intervene. Sproat resigned in March 1880, to be replaced by Peter O'Reilly.

3
Exclusive Fisheries

The processes that began or were renewed in the 1870s – the rise of an industrial commercial fishery, the introduction of Canadian fisheries law, the growing non-Native presence on the land, and the allotment of Indian reserves – came into clearer focus in the 1880s. As they did, the contradictions between them raised tensions: there would be Indian reserves, but no recognition of Native title; the allotment of reserve land would follow rather than precede non-Native settlement; many reserves would be designated as fishing stations, but without legal protection for the fisheries; the industrial commercial fisheries needed labour, but not necessarily Native labour. However, in 1881, when Peter O'Reilly assumed the position of Indian reserve commissioner, the consequences of these contradictions were not yet fully apparent. His work, particularly his attempt to reserve exclusive Native fisheries along the Fraser and Nass rivers, would throw some of them into sharp relief. In fact, O'Reilly's marking out of exclusive fisheries precipitated a debate between the Dominion departments of Fisheries and Indian Affairs that would last nearly twenty years. Before turning to that debate and its legal underpinnings in the following chapter, this chapter focuses on its subject – Reserve Commissioner O'Reilly's mapping of exclusive Native fishing rights. The chapter also considers the recognition of Native fishing rights inherent in the contractual arrangement between a Tsimshian chief and two canning companies on the north coast in the late 1870s.

Peter O'Reilly was a county court judge, although without formal legal training, who briefly served as reserve commissioner in 1870 before the process ground to a halt after Confederation. He lived in a stately Victoria home, was good friends with the lieutenant-governor Joseph Trutch, and was married to Trutch's sister. In short, he was well connected to the provincial elite, whose vision of an appropriate Native land policy had prevailed in British Columbia. It was a vision with which O'Reilly agreed, and he was a reserve commissioner the provincial government, at least, could trust to represent its interests.[1] It is all the more surprising, therefore, that it was O'Reilly's approach to the fisheries that provoked the greatest uproar. But then, fisheries were a Dominion responsibility, and it was the Dominion Department of Marine and Fisheries (Fisheries) that would be most concerned about his work. The province had yet to take much of an interest in fish.

The provincial government did not provide new instructions to the incoming reserve commissioner, although, given O'Reilly's connections, the informal channels were probably more than adequate for the province to convey its intentions. For its part, Indian Affairs told O'Reilly, as it had told the earlier commissioners, that in allotting reserves he was to have "special regard to the habits, wants and pursuits of the Band," and that he should "interfere as little as possible with any fur trading posts, settlements, clearings, burial places and *fishing stations* occupied by them to which they may be specially attached." He was, moreover, to pay special attention to their fisheries: "Their *fishing stations* should be very clearly defined by you in your reports to the Dept. and distinctly explained to the Indians interested therein so as to avoid further future misunderstanding on this most important point."[2] Given the growing unrest over access to fishing grounds where the canneries were active, and a history of conflict over fisheries in Ontario, Indian Affairs wanted the reserve commissioner to secure Native access to the fisheries. This, apparently, is how O'Reilly understood his instructions.

Exclusive Fisheries on the Fraser River, 1881

The regular, predictable migration of Pacific salmon up the Fraser River drew people to the river. "Salmon concentrate energy in time and space," writes anthropologist Michael Kew, and this "is why they are so beneficial to humans."[3] At certain places the fat content of the salmon, the run of the river, the local climate, and the particular human culture created focal points where salmon could be harvested and preserved in great quantity. One of these places was the stretch of river above Yale at the start of the Fraser Canyon.[4] Another was at the mouth of the Bridge River, just north of Lillooet in Stl'átl'imc territory along the Fraser River. Thousands of people gathered along these stretches of the river to catch and dry great quantities of sockeye salmon as they migrated upstream. Earlier runs of chinook salmon were also important.

The communities that depended on these fisheries regulated the access to them. Anthropologist Steven Romanoff posits a division of modes of ownership. An individual might own a certain location identified in time

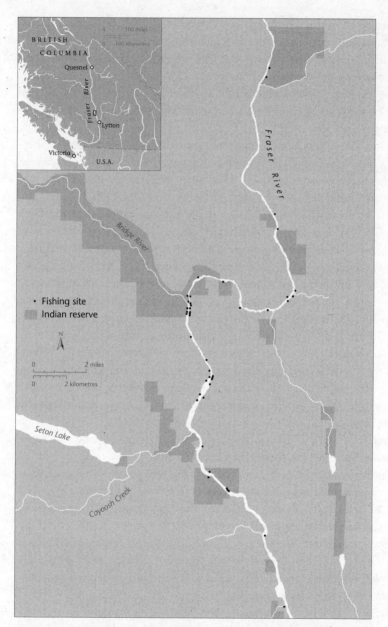

Figure 3.1 Stl'átl'imc fishing sites along the middle Fraser. Note the particular concentration of sites at the mouth of the Bridge River. Most of the sites identified particular rocks from which nets could be dipped or set. Some of the sites included several fishing locations, all within a few metres of each other. | *Source:* Adapted from Dorothy I.D. Kennedy and Randy Bouchard, "*Stl'átl'imx* (Fraser River Lillooet) Fishing," in *A Complex Culture of the British Columbia Plateau: Traditional* Stl'átl'imx *Resource Use,* ed. Brian Hayden, 321-26 (Vancouver: UBC Press, 1992).

and space so as to allow fishing from a particular rock for a specified species of salmon at a certain time of year. These rights could be inherited. Other locations, usually described as a stretch of the river, were associated with neighbouring residence groups who controlled access. Finally, some locations were public, open to all Stl'átl'imc as well as to those peoples with whom they had reciprocal or trading relationships. These "public" sites were generally the most productive.[5] Anthropologists Dorothy Kennedy and Randy Bouchard have mapped some of the locations that were individually "owned" (Figure 3.1).[6] "Ownership," they suggest, seldom amounted to exclusive rights for individuals or groups, but it did confer the right to regulate access.

In 1881, his first season in the field as reserve commissioner, O'Reilly focused on the peoples who lived along the middle stretches of the Fraser River between Lytton and Quesnel. More than twenty years after the prospect of gold had brought thousands of miners to the Fraser, the Stl'átl'imc and Secwepemc peoples still had no reserves. Once in the field, O'Reilly encountered the same difficulties that Sproat had faced in the interior – inadequate land and water to create viable agrarian economies. Like his predecessors, O'Reilly was looking for agricultural land, but there was little enough in the dry and mountainous interior, and the province had alienated most of what there was, including some of the land and water rights that had been allotted as reserves under Governor Douglas, to settlers.[7] O'Reilly did not find or allot much farmland, but he did attempt to protect the Native fisheries.

The reserve commissioner worked quickly and, from the province's perspective, efficiently. He talked with those Native peoples who happened to be present, seldom going out of his way to consult, and he allotted reserve land based on information conveyed in brief meetings. Figure 3.2 shows O'Reilly's route and all the reserves he allotted during this first circuit. Among other things, it reveals the importance of fisheries in what is a mountainous and dry country (Photo 3.1). With a few exceptions (primarily graveyards and a few plots of arable land with access to water), the reserves bordered or straddled rivers or lakes and were intended to secure access to fish.

O'Reilly began in Secwepemc territory at Williams Lake, where he allotted six reserves and protected a collection of gravesites on alienated land. The six reserves included a four-thousand-acre allotment at the head of Williams Lake and two much smaller hay meadows. These were among the few reserves that held some agricultural potential. The other three reserves were fishing stations: one near the outflow of Williams Lake, and two on the Fraser at the mouths of San Jose and Chimney creeks. At Soda Creek,

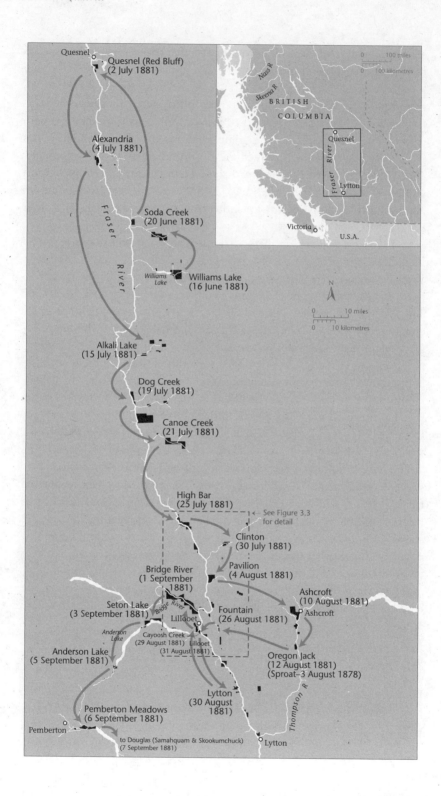

Quesnel
Quesnel (Red Bluff)
(2 July 1881)

Alexandria
(4 July 1881)

Soda Creek
(20 June 1881)

Fraser River

Williams Lake
Williams Lake
(16 June 1881)

Alkali Lake
(15 July 1881)

Dog Creek
(19 July 1881)

Canoe Creek
(21 July 1881)

High Bar
(25 July 1881)

See Figure 3.3
for detail

Clinton
(30 July 1881)

Pavilion
(4 August 1881)

Bridge River
(1 September 1881)

Ashcroft
(10 August 1881)
Ashcroft

Seton Lake
(3 September 1881)

Bridge River

Fountain
(26 August 1881)

Anderson Lake

Lillooet

Cayoosh Creek
(29 August 1881)   Lillooet
(31 August 1881)

Anderson Lake
(5 September 1881)

Oregon Jack
(12 August 1881)
(Sproat–3 August 1878)

Thompson R.

Lytton
(30 August 1881)

Pemberton Meadows
(6 September 1881)

Pemberton

to Douglas (Samahquam & Skookumchuck)
(7 September 1881)

Lytton

N

0   10 miles
0   10 kilometres

Inset map:

Nass R.
Skeena R.
BRITISH
COLUMBIA

Fraser River

Quesnel

Lytton

Victoria
U.S.A.

0   100 miles
0   100 kilometres

◄ *Figure 3.2*   Peter O'Reilly's first circuit as Indian reserve commissioner, beginning 16 June 1881 at Williams Lake.

Photo 3.1   The Fraser River north of Lillooet. In this dry, mountainous terrain, the Fraser River salmon were the primary source of sustenance and wealth for Native communities. | *BC Archives A-04679*

O'Reilly's next stop, he allotted two reserves, one of which was a fishing station on the Fraser. He set aside a similar fishing station for the Quesnel and reserved a few acres for them at the outflow of a small lake where they harvested fish through the winter.

O'Reilly then headed south, down the Fraser, allotting similar reserves for fishing purposes, but he now included the phrase "the exclusive right of fishing" in his description of many of the reserves. For the Alexandria, O'Reilly set aside "the exclusive right of fishing on the West bank of the Fraser river" for the entire length of Reserve No. 3, a 1,200-acre parcel that bordered on the river.[8] For the people at Alkali Lake, O'Reilly marked out "the exclusive right to fish on the left [east] bank of the Fraser river, from the mouth of the Chilcotin river to the mouth of Little Dog Creek, an approximate distance of 4 miles," and set apart "a fishery reserve situated on the North shore of Lac la Hache."[9] He also identified the right of the Canoe Creek band to fish in a small lake and reserved 5.5 miles of the Fraser as their exclusive fishery, even though none of their reserves bordered on the river.[10]

O'Reilly's allotment of reserves and associated fisheries in the next section of the Fraser is depicted in Figure 3.3. The people at High Bar secured a six-mile tract of exclusive fishing on both banks of the Fraser, the length of their long and narrow 2,600-acre reserve, and the Clinton, a similar stretch of river immediately below that.[11] For the Pavilion, O'Reilly set aside an exclusive fishery spanning the Fraser and extending from Leon Creek, which bisected one reserve, south to a point just above Gibbs Creek (11 Mile Creek), a distance of fifteen miles and the longest stretch of exclusive fishing rights to any one group. After a detour to the Thompson River, where he allotted reserves and exclusive fisheries, O'Reilly returned to the Fraser, entering Stl'átl'imc territory and moving quickly through it. Just north of the Bridge River, he marked out an exclusive fishery on both sides of the Fraser for the Fountain, a distance of 4.5 miles.[12] To the south, for the Cayoosh Creek band, he noted an exclusive fishery on the south bank of Cayoosh Creek and then south along the west bank of the Fraser. O'Reilly reserved to the Lillooet the east bank of the Fraser opposite Cayoosh Creek and another four-mile stretch on both sides of the river from Cayoosh Creek to a point near the mouth of the Bridge River. He also reserved their exclusive fishery on Seton Creek at the outflow of Seton Lake.[13] To conclude the intricate division of the fishery on the Fraser near the Bridge River, O'Reilly set aside for the Bridge River band a three-mile stretch of exclusive fishing that ran from the confluence north to the Fountain fishery.

O'Reilly marked off the reserves and divided the Stl'átl'imc fisheries near the mouth of the Bridge River in less than a week. In this short

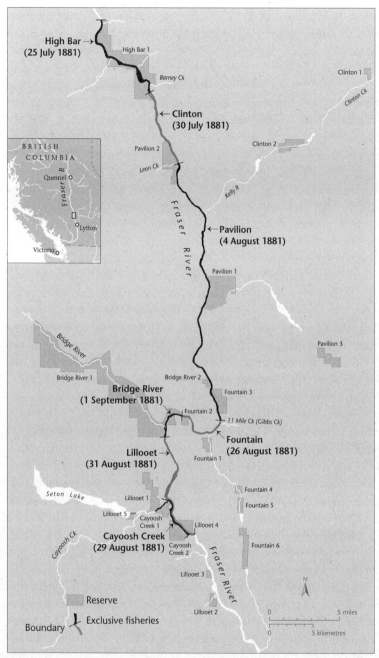

Figure 3.3 Exclusive fisheries reserved along the Fraser River. Note the long stretch of river that O'Reilly set aside for the Pavilion. South of Cayoosh Creek, the Lillooet had exclusive fishing rights on the east bank of the Fraser, the Cayoosh Creek on the west bank.

window, his division of the fishery could, at best, only roughly approximate the closely owned traditional rights of access that were seasonally and technologically specific. He was moving much too quickly to integrate this level of detail, nor was he particularly interested in doing so. Even Sproat, who had moved much more slowly and thoroughly, and in some cases had attempted to micromanage compromises between the interests of settlers and Natives, had not tried to replicate the detailed patterns of Native ownership. O'Reilly's work left some inconsistencies. The Bridge River Reserve No. 2, for example, bordered the Fraser where O'Reilly had reserved an exclusive fishery to the Pavilion. Moreover, his designation of exclusive fisheries, had they been confirmed, might well have entrenched boundaries that were much more fluid in practice. Nonetheless, Kennedy and Bouchard suggest that O'Reilly's division of the fisheries on the Fraser corresponds to the traditional fishing territories of the local villages or resident groups.[14]

When he had finished working on the Fraser, O'Reilly had reserved exclusive fishing rights for approximately forty-five miles of the middle Fraser, an area that contained some of the most productive fisheries in the province. From there, he moved south and west, reserving land and exclusive fisheries to the bands at Seton Lake and Anderson Lake, the Mount Currie band near Pemberton, and the Douglas band on the Lillooet River. Indian Affairs forwarded the minutes of decision, which contained descriptions of all these fisheries, to the Department of Fisheries.

Fishing Contracts and Fishing Reserves on the North Coast

In October 1881, in response to a growing conflict between cannery operators and Native peoples over important fishing stations, O'Reilly went to British Columbia's north coast to allocate reserves. Nisga'a chiefs blamed the two canneries, which had opened near the mouth of the Nass River that year, for a diminished salmon run and were concerned that more canneries and boats would be operating the following year. They sought greater regulation of the non-Native fishery and asked that the government restrict the number of boats.[15] Indian Superintendent I.W. Powell forwarded the Nisga'a petition to Indian Affairs, indicating that he thought it was "of vital necessity that fishing stations and certain streams should be specially reserved for them just as agricultural lands are set aside in the Interior."[16] Some months later, he recommended that O'Reilly reserve exclusive Native fisheries near the mouth of the Skeena, and he indicated that the inspector of Fisheries, A.C. Anderson, supported him.[17] Indian Affairs officials in Ottawa concurred, forwarding these views to the Department of Fisheries.[18]

The Metlakatla Tsimshian at Port Simpson wrote directly to O'Reilly, expressing concern about the size of their reserves and access to their fisheries: "Independent of our land reserve we desire that our fishing stations on the Nass and Skeena Rivers be secured to us as a matter of greatest importance."[19] This was hardly a surprising request. The sense of ownership among northern tribes was strong, something that the early white settlers had encountered. Methodist missionary A.E. Green, who arrived on the Nass River in 1877, had sought to build a mission but gave up for lack of permission to build on Native land. "The idea of ownership," he noted, "was so strong among the Indians ... Every mountain, every valley, every stream was named, and every piece belonged to some particular family." When the Methodists eventually did build a mission, the site, according to O'Reilly, "unfortunately was badly chosen, being surrounded by mountains of rock with no land of value nor any fisheries in the immediate neighbourhood."[20] Perhaps this was the only land on which the Methodists had been allowed to build. Green went on to say that Native ownership "was recognized by all the white men, viz. Harvey Snow, James Grey, J.J. Robinson, who rented small sites from the Indians for fishing purposes, and paid the Indians regular rent for the same."[21]

Several examples of these early contracts survive. In June 1877, C.S. Windsor signed a contract with Ts'ibasaa (Paul Sebassah), a leading Kitkatla chief,[22] to secure access to a fishery:

Agreement made this twenty-third day of June, one thousand eight hundred and seventy seven, between C.S. Windsor of Inverness British Columbia, on the one part And Paul Sebassah Chief of Kathlahtla on the other part.

Whereas the said Paul Sebassah agrees to give to the said C.S. Windsor, the exclusive privilege to fish for salmon in the Kitsumalarn River from year to year for from one to four years, and not longer unless another agreement should be made between both parties, reserving the right only to himself and brother, and to allow the said C.S. Windsor to erect a cannery and any other buildings that he may think well to erect.

In consideration of which the said C.S. Windsor agrees to pay to the said Paul Sebassah, one hundred dollars per year, one half to be paid in cash and the other half in goods.

The [rent to?] commence from the time the said C.S. Windsor commences to erect the buildings.

In witness hereof the parties to these presents have hereunto set their hands and seals, the day and year first above written.

C.S. Windsor

Paul Sebassah ✗ his mark[23]

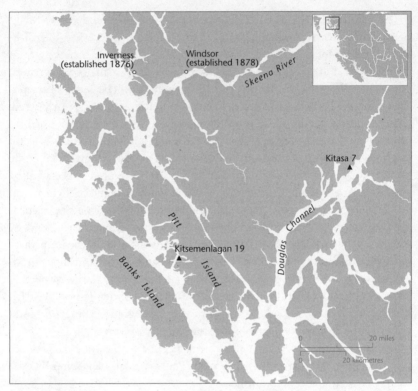

Figure 3.4   Skeena canneries and location of fishing contracts. This map shows · canneries on the Skeena in 1878, as well as Indian reserves that correspond to the possible locations of two contracts in which Ts'ibasaa (Paul Sebassah), a Kitkatla chief, granted exclusive fishing rights in 1877 and 1878.

Ts'ibasaa agreed to transfer the "exclusive privilege" to fish for salmon in the Kitsumalarn River and the right to build a cannery and other buildings nearby. The location appears to correspond with what became the Kitkatla reserve Kitsemenlagan 19 in Curtis Inlet on the west coast of Pitt Island (Figure 3.4). In return, Windsor contracted to pay $100 per year, once he began building the cannery, with an option to renew for up to four years. Ts'ibasaa retained his right and that of his brother to fish in the river, and it seems likely that they would have planned to sell their catch to Windsor's cannery. However, in 1878, Windsor built a cannery, which became the Aberdeen cannery, near the mouth of the Skeena River, more than sixty miles away from the fishery that he had leased from Ts'ibasaa. Although it seems unlikely that Windsor paid any rent, the fact that he felt compelled to enter such an agreement reveals the extent to which the Kitkatla retained control over their traditional territories. In this context, the expedient response for commercial capital was to contract for access to the fishery.[24]

In 1878, Ts'ibasaa entered another contract with the North West Commercial Company. This company had opened the first cannery on the north coast, the Inverness cannery at Port Edward near the mouth of the Skeena, in 1876. Three years later it appeared to be securing its supply of fish by agreeing to the following terms with Ts'ibasaa for the salmon fishery at Kitsah:

Metlakatlah, 14th June /78

I the undersigned being the proprietor of the salmon fishery at Kitsah hereby sell the exclusive right of fishing in said waters for the present season to the N.W. Com. Co. at the price of fifty dollars of which twenty five dollars at the beginning of the season and twenty five at the end of it – payable half in goods and half in cash – and in case of the said Company being willing to take the said fishery from year to year afterwards I engage to let them have it at whatever price for each season I may be offered by any other party – and I guarantee that no one else but the men employed by the Company will be allowed to fish with the exception of my own family for their own food.

his

(*Signed*)   Paul ✗ Sebassah

mark

We accept the above contract and agree to be bound by it and carry out its provisions.

For the N.W. Comm. Co.

(*Signed*)   W.M. Neill[25]

It was a somewhat different contract from that signed with Windsor. The cost of the Kitsah lease was half what Windsor had agreed to, perhaps reflecting the lower value of the fishery or the fact that the North West Commercial Company had been operating in the region for three years and knew more about its value than did Windsor. There was also no provision for building a cannery, further suggesting that the fishery was not productive enough to warrant a processing plant on its doorstep or that the company did not wish to build another cannery. The lease would continue year to year, with an option for the company to match any competing offer. Ts'ibasaa, whose ownership of the fishery was established in the contract, also assumed the responsibility for ensuring that the company enjoyed an exclusive fishery, reserving the right for his family to continue their food fisheries. They could not, under the terms of the contract, fish commercially. The location of the lease is unclear, but it may correspond to what would become the Kitamaat reserve on Douglas Channel, Kitasa 7. If so, it was several hundred kilometres by boat to the company's cannery near the mouth of the Skeena.

These types of leases were common in Upper Canada in the eighteenth and nineteenth centuries.[26] Victor Lytwyn has provided details of a series of leases granted by the Saugeen Ojibway to non-Native fishers in the 1830s and 40s.[27] It seems, furthermore, that as late as the 1830s the Crown issued licences of occupation to the non-Native lessees, confirming their right to an exclusive fishery under the lease. In other cases, the Crown leased fisheries to non-Native fishers on the understanding that they come to some arrangement with the Native owners of the fisheries as well. There is no evidence that the Crown confirmed the leases in British Columbia, or paid any attention to them at all, and the practice of leasing fishing grounds directly from Native peoples appears to have ended in the 1880s, when Dominion officials asserted that the common-law doctrine of the public right to fish precluded exclusive fisheries in tidal waters (see Chapter 4).

When the Metlakatla Tsimshian appealed to the reserve commissioner to set aside reserves in 1881, it was not out of a newly found sense of proprietorship, but rather from a long-established tradition of ownership. What was new was the appeal to the reserve commissioner, and hence to the state, for assistance in protecting property rights. The petitioners sought assistance from one arm of the state – the reserve commission and, to some extent, Indian Affairs – to help manage the canneries that had the backing of other arms – the Dominion Department of Fisheries and the provincial government. O'Reilly allotted eleven reserves to the Metlakatla Tsimshian in October 1881, ten of which were intended primarily to secure access to fisheries. Reserve No. 5 included "the exclusive right of fishing in 'Cloyah' river for a distance of 1½ miles from its mouth."[28] He described other reserves as "fishing stations" or noted their association to nearby fisheries.

Earlier in October, O'Reilly had been up the Nass River allotting reserves and allocating exclusive fisheries to the Nisga'a. These are mapped in Figure 3.5. O'Reilly considered the Nisga'a claims that the canneries were ruining the salmon runs to be groundless, but of the fifteen reserves he set out between Aiyansh and Nass Bay, eight included the exclusive right of fishing in the Nass River or its tributaries.[29] The reserve at Aiyansh, for example, included "the exclusive right of fishing in the Nass River the entire length of this reserve and also in Chemanuc Creek." Most protected fisheries were in waters adjacent to the reserve, but at Reserve No. 2, O'Reilly included "the exclusive right of fishing in the Nass River for a distance of 2 miles upstream from this Reserve." As O'Reilly worked his way down the Nass, he continued to recognize exclusive fisheries until Reserve No. 9, near the tidal boundary, where he indicated that adjacent fisheries in the Nass were "reserved for the use of the Indians."[30] Reserve No. 13 included the exclusive right of fishing in a small tributary. Four others were intended to secure access to salmon or oolichan fisheries. In his field minutes, written almost six months after his visit to the north coast, O'Reilly claimed to have informed the Tsimshian and Nisga'a that their fishing was subject to the laws of the Dominion and "that in assigning to them the several stations on the Coast, and tidal waters, *no* exclusive right of fishing was conveyed."[31]

These reserves were so hastily allotted and ill considered that they produced extended conflict between the Tsimshian and Nisga'a, and within the Nisga'a community itself.[32] According to Methodist missionary T. Crosby, "Mr. O'Reilly, Land Commissioner, came up in 1881, without giving notice of his coming. He called together a few people who were at home. They complained that they had not had any notice, and hence so few of their people were at home, but went on to explain what lands they wished to keep for their people. He pooh-poohed and said he could not listen to such long speeches, he must get on with his work."[33] As a result, insufficient land had been allotted. In 1887, six years after O'Reilly's visit, a delegation of Nisga'a and Tsimshian travelled to Victoria to tell the provincial government of their dissatisfaction with the reserves and also with the lack of a treaty.[34] Premier Smithe curtly dismissed the idea of a treaty but indicated, somewhat disingenuously, that the provincial government might consent to additional reserves if they were shown to be necessary. He clearly did not think they were necessary, and O'Reilly also believed that he had allotted an adequate land base. Responding to the claims of Nisga'a chief John Wesley that every chief had a hunting territory and fishing ground that should be protected, O'Reilly explained what he had done:

> They [Nisga'a chiefs] each have a little spot which they are in the habit of calling their own. Every inlet is claimed by some one, and were I to include all these, it would virtually declare the whole country a reserve; this arrangement I could not justify. To lay out all the inlets pointed out and claimed by them, would be impossible. *They were given the right to all streams which run through their reserves, and every fishing ground pointed out by them, of every sort or kind, was reserved for them.* There was no difficulty in doing this, as the fish of special value to the Indians the white men do not care for, therefore their interests do not clash. But to declare every inlet, nook, and stream an Indian reserve would virtually be to declare the whole country a reserve.[35]

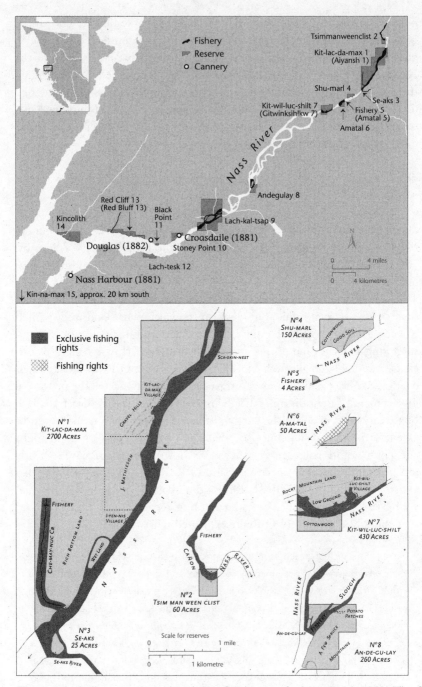

Figure 3.5 Indian reserves and exclusive fisheries on the Nass River, 1881. The first map displays the Indian reserves and exclusive fisheries allotted by O'Reilly on the Nass River in October 1881. The lower map and facing map are based on O'Reilly's

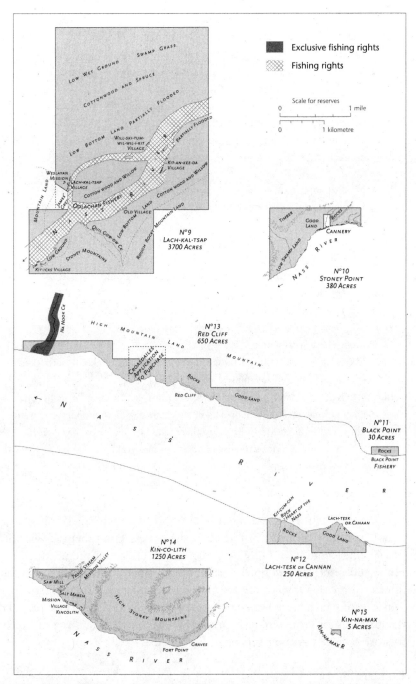

field sketches of these reserves, with additional shading to reveal the fishing rights. Note the cannery in the middle of Reserve No. 10. | *Source:* O'Reilly, Sketches accompanying Minutes of Decision, 20 October 1881, Fed. Col., vol 9.

Taken at his word, O'Reilly had allotted not just land but also portions of rivers and distinct fishing grounds. In 1887 he challenged the Nisga'a and Tsimshian, asking them if they could "name any fishery that has been omitted? Any that has not been laid out for them?" Richard Wilson, a Tsimshian chief, responded that they did not have the maps to know exactly what had been set aside as reserve, and in fact the reserves had only just been surveyed. But by this point in the meeting it was clear that the chiefs had come to talk about a treaty, while the government officials were only prepared to tinker with the existing reserves.

> Hon. Mr. Robson [Provincial Secretary]: When you [O'Reilly] were up there you laid out all the fishing grounds and reserved them?
>
> O'Reilly: Every one.
>
> Hon. Mr. Robson: Did you miss any? Because, surely, if you did, it is a grievance which could be easily remedied.
>
> Dr. Powell [Indian Superintendent]: I think the grievance they have is, that they want a treaty.
>
> Hon. Mr. Davie [Attorney General]: That is what they want.
>
> Hon. Mr. Smithe [Premier]: They are simply misguided.
>
> Dr. Powell: There is no doubt of it.
>
> Mr. O'Reilly (to Burton): Was anything left out? Was there a single cultivated patch of land, which the Indians shewed us, that was omitted?
>
> Burton [Nisga'a translator]: They do not wish to answer or argue. We are here just to tell you what we want. Then you will look it over yourselves, and if you want to know further send Mr. O'Reilly up there, and he will talk to the chiefs himself. There is no use for us to answer questions and argue with Judge O'Reilly here, or with the Government. We are just merely sent to tell you what we want.[36]

It was clear that the disagreement was too great to resolve. But it is also clear that fishing grounds, not just land associated with fisheries, were the crucial consideration in O'Reilly's reserve allocations on the north coast. He would return to the north coast in September 1888 to allot another fifteen reserves in Observatory Inlet (which he had missed on his first visit), fourteen of which were intended to secure access to a fishery or provide the materials for processing fish, primarily cedar for making boxes in which to store oolichan grease (Appendix, Figure A3).

Conclusion

After two circuits, one in the central interior and another on the north coast, O'Reilly had established his modus operandi as reserve commissioner.

While emulating the earlier commissioners in many respects, his methods also marked a significant departure. He travelled much more quickly than the Joint Indian Reserve Commission or Sproat, and he had no patience to work out the intricate compromises that his predecessor had sought to formalize through reserve allotments. Nonetheless, he pursued a similar goal of protecting Native fisheries, and his methods, while less nuanced, were more transparent. He set aside exclusive Native fisheries, most often in waters adjacent to reserve land, but occasionally extending well beyond the reserve boundaries. This included approximately forty-five miles of exclusive fisheries on the middle Fraser, which he allocated to the Secwepemc and Stl'átl'imc. On the Nass River, O'Reilly set aside the important fishing grounds, not just the land associated with those grounds. The fact that he was not moving slowly or carefully enough to make sure that he allocated the fisheries accurately does not diminish his intention, acting on his instructions, to establish the waters in which Native people were to have exclusive fishing rights. It was this transparency, perhaps, that provoked a strong reaction from the Department of Fisheries. These exclusive fisheries, the department would argue, violated the public's right to fish.

4
Exclusive Fisheries and the Public Right to Fish

The public right to fish is a doctrine of the common law. As such, it is a creature of judicial interpretation, operating to limit the capacity of the Crown to grant exclusive fisheries and, in doing so, protecting the right of the public to fish. It is, of course, but one doctrine among the panoply that constitute the common law. This was the body of law – judge-made and statutory – that arrived with English-speaking settlement in North America and throughout the British Empire. In colonies that were settled, wrote William Blackstone in his *Commentaries on the Laws of England* (1765-69), "the law is the birthright of every subject, so wherever they go they carry their laws with them."[1] The introduction of English law with British settlement was followed by its formal reception, the date of which delimited the moment at which further statutory enactments in England would not apply to the colony. Judicial pronouncements from England might still be relevant, and English courts might be the courts of last resort, as they were in Canada until 1949, but the English statutory foundation of a colony was set by the date of reception. Further statutory developments would come from within the colony itself.[2] In British Columbia, the common law was formally received by statute with the founding of the mainland colony of British Columbia in 1858.[3]

Although the English common law accompanied British settlement, the settlers' birthright was not to the common law as it existed in England. Blackstone set out its limits in a subsequent edition of his *Commentaries*: "But this must be understood with very many and very great restrictions. Such colonists carry with them only so much of the English law, as is applicable to their own situation and the condition of an infant colony."[4] In British Columbia, this doctrine was codified in statute: the civil and criminal laws of England applied in the colony "so far as the same are not from local circumstances inapplicable."[5] Determinations of applicability might be statutorily decreed or they might be left to judges, but in either event the transmission of English law involved the parsing of individual doctrine for its suitability to the local. Property scholar Bruce Ziff has described this as the "crucible of *applicability*," a process in which English law was measured for its colonial fit.[6]

The public right to fish provides a marvellous example of this crucible of applicability. Judges in British North America grappled with a doctrine that had developed in the different geographical circumstances of the British Isles. Did the public right to fish apply in the vast inland lakes and waterways, of which there were no counterparts where the doctrine had emerged? The answer had consequences. It would affect the legality of the exclusive fisheries that Indian Reserve Commissioner Peter O'Reilly marked off in 1881. Beyond the particular application of the public right to fish, its use and, I will argue, misuse in the debate over the legality of the exclusive fisheries set aside by O'Reilly raises interesting questions about the role of state law in the processes of dispossession that mark the colonial encounter.

O'Reilly's allocation of exclusive fishing rights to Native peoples on the Fraser and Nass rivers in 1881 provoked controversy almost immediately. Officials within the Dominion Department of Marine and Fisheries (Fisheries) demanded to know on what authority O'Reilly had presumed to grant exclusive fisheries in addition to the Indian reserves. O'Reilly, they argued, could only recommend parcels of land that should be set aside as reserves, not exclusive fisheries that were contrary to the public's right to fish.

This chapter opens with a description of that right and the challenges in its transmission to North America. It then turns to the reactions against and in support of O'Reilly's allocation of exclusive Native fisheries. Finally, it questions the legal foundation of the opposition to O'Reilly's work and suggests that O'Reilly was working carefully and securely within the bounds of Canadian law when he marked off exclusive Native fisheries in non-tidal waters.

Lingering in the background, and addressed at the end of this chapter, is the larger question of how best to characterize what O'Reilly was doing. Was he recognizing rights to exclusive fisheries that already existed? Or was he, as agent of the Crown, granting rights to exclusive fisheries? Opposition to his work was predicated on the assumption that he was granting and therefore creating rights rather than recognizing existing rights. This view – that rights to land and fish emanated from the Crown – came to prevail in the late nineteenth century. It was part of the consolidation and centralization of power in the colonial state. It was not, however, the only view. Native peoples stated repeatedly that the sources of their fishing rights lay in their legal traditions, not in the Crown. In this view, a provision such as the fisheries clause in the Douglas Treaties did not create rights but, instead, converted existing rights into treaty rights (see the discussion in Chapter 1). These prior rights of Native peoples to their fisheries, therefore, might be confirmed by the work of the reserve commission, but they were not created by it. This interpretation of Native fishing rights remained in the background in British Columbia along with the issue of Native title, never disappearing but pushed well into the shadows by the prevailing view that rights to land and fish originated from the colonial state. This latter view was less a part of an overt strategy of colonial control than it was simply presumed. Native peoples countered this supposition as best they could, but their pleas for recognition of an alternative source for legal rights received little attention in the late nineteenth and early twentieth centuries.

However one characterizes O'Reilly's work as reserve commissioner, by 1898, when he retired, Native peoples' access to their fisheries had been severed from the reserve grants. Even reserves that were in close proximity to local fisheries, and that had been allotted specifically to secure access to those fisheries, no longer conferred any rights of access to fish, far less exclusive fisheries. The common-law doctrine of the public right to fish, as interpreted by senior Fisheries officials in the late nineteenth century, effectively opened many locally owned and managed fisheries to outsiders, in most cases reallocating the resource from Native to non-Native fishers.

The Public Right to Fish

The English common law did not recognize property in fish until they were caught, but it did recognize a property interest in the right to catch fish. It did so by creating a presumption that the owner of the soil underlying a body of water – the *solum* – had, as an incidence of the ownership of the *solum*, an exclusive right to fish in the water above. This exclusive right to fish could be severed from the ownership of the underlying bed, but severing the fishery from the larger property interest had to be done explicitly in a grant or other instrument used to transfer the interest in property; it would not be presumed.

In determining who owned the *solum*, and therefore the right to fish in waters above it, the common law created another presumption that those who owned the land adjacent to a defined body of water, known as riparian owners, also owned the *solum* to the midpoint of that body, a doctrine known as *ad medium filum aquae*. Riparian owners whose property surrounded a body of water owned the entire bed and the exclusive right of fishing in the entire body of water as a result. In England, this presumption applied to all non-navigable waters, understood in law to mean non-tidal waters.

In tidal waters, the Crown owned the fisheries. It is not clear whether this right arose as an incidence of the Crown's underlying interest in the tidal foreshore or by some other means, but the Crown held the right of fishing in tidal waters, which English lord chief justice Mathew Hale described in his much-cited passage as "the sea or creekes or armes thereof."[7] In contrast to the situation in non-tidal waters, however, the Crown's ownership of the fisheries in tidal waters was subject to the common-law doctrine of the public right to fish. In effect, the Crown held the right to fish in tidal waters in trust for the public. This prevented it from granting exclusive fisheries in tidal waters to individuals or corporations, something that it and others who held the right of fishing in non-tidal waters were free to do. Parliament could authorize the Crown to grant exclusive fisheries in tidal waters, but such grants were not part of the Crown's prerogative.

The origin of the public right to fish is often said to be the Magna Carta (1215), and that may be so, although the language used is far from explicit. It seems more likely that the thirteenth-century charter of rights, which the barons forced upon King John, became in the eighteenth and nineteenth centuries a proxy for the long-standing custom of public access to the fisheries in tidal waters.[8] The uncertainties of origin notwithstanding, Magna Carta became the touchstone for the public right to fish in the common law, creating a temporal boundary delimiting a time when the Crown had within its prerogative the right to allocate exclusive fisheries in tidal waters (before Magna Carta) from a time when it did not (ever since).

That the public right to fish applied only in tidal waters worked well in England, where there were no large non-tidal waterways or lakes to complicate matters. Some of the large lakes in Scotland and Ireland presented a few difficulties, but the courts maintained a firm line in the cases that arose: only navigable waters, defined in law as tidal waters, were subject to the public right to fish.[9] Circumstances in North America were considerably different. The vast inland lakes and river systems that were navigable but non-tidal required the rethinking of a body of law that had emerged in a context where tidal and navigable were sufficiently synonymous in fact to be considered so in law. Did the *ad medium filum aquae* rule create a presumption that the owner of a small lot beside one of the great lakes also owned the *solum* to the middle of the lake and, therefore, held an exclusive right of fishing in the waters above that land? And if in these circumstances the common law were modified to create a presumption that the Crown retained ownership of the *solum* and thus the fishery, was the Crown's interest subject to the public right to fish? These became important questions in British Columbia when the reserve commissioners allotted land in the 1870s and 1880s to secure Native access to fish.

Exclusive Native Fisheries and the Public Right to Fish

Within days of receiving copies of Peter O'Reilly's minutes of decision creating the reserves along the Nass River and setting aside exclusive fisheries, the Dominion commissioner of Fisheries, W.F. Whitcher, wrote to the Department of Indian Affairs (Indian Affairs) demanding to know "under what authority Mr. O'Reilly has appropriated public fishing privileges in public waters not even connected with Indian land Reserves – for the exclusive right of Indians."[10] Indian Affairs had sent copies of the minutes of decision creating exclusive fisheries along the middle Fraser, but few if any non-Natives were fishing in these waters, the canneries near the mouth of the Fraser had prior access to salmon, and Fisheries had not responded. The exclusive fisheries on the Nass, however, were much closer to, and might damage the viability of, the emerging canning industry on the north coast. This, it appears, provoked the immediate response from Whitcher. He believed that Fisheries had the sole authority, under the *Fisheries Act*, to grant exclusive fisheries in public waters, understood to include tidal and navigable waters.[11]

All of the exclusive fisheries that O'Reilly had designated were in non-tidal waters, but most were in waters that Fisheries considered navigable. To Whitcher's demand for an explanation, John A. Macdonald, prime minister and superintendent general of Indian Affairs, replied that he had "considered it expedient and proper to instruct him [O'Reilly], while engaged in assigning these lands, to mark off the fishing grounds which should be kept for the exclusive use of the Indians and he is following those instructions."[12] In turn, the minister of Fisheries, A.W. McLean, replied that the Crown's prerogative was limited, that it could only grant exclusive fisheries if authority had been delegated from parliament, that parliament had delegated this authority to his department under the *Fisheries Act*, and that Fisheries had no intention of limiting non-Native access by granting exclusive Native fisheries:

> Fishing rights in public waters cannot be made exclusive excepting under the express sanction of Parliament, and ... Indians are entitled to use the public fisheries only on the same conditions as white men, subject to the *Fisheries Act* and Fishery Regulations. The mere assignment of these fishery privileges by Indian Agents, or the abstention of this Department from otherwise disposing of them – which there was no intention to do pending careful consideration of all the circumstances of each case – could not legally exclude the public from fishing therein.[13]

Indian Affairs, on the other hand, sought to protect Native fishing grounds and had instructed the reserve commissioner to include them within the reserve allotments. Behind this position lay three considerations. First, Native peoples in British Columbia depended heavily on their fisheries. Reallocating their fisheries would cause great hardship and, from Indian Affairs' perspective, would increase the number of dependents that it would be responsible to support. Second, from its experience in the Great Lakes region, Indian Affairs knew that Native peoples would defend their fisheries tenaciously, and it did not want to provoke unrest in a volatile situation.[14] Finally, the unresolved question of Native title and fishing rights lingered in the background. Officials at Indian Affairs were well aware that they were trying to resolve the land question in British Columbia without treaty at the same time that they were negotiating treaties in what would become the Prairie provinces. The absence of treaties did not mean an absence of rights, including rights to established fisheries. This was not lost on former reserve commissioner Gilbert Malcolm Sproat, who had pointed to the Douglas Treaties as evidence of fishing rights across British Columbia, or on Prime Minister Macdonald, who had written that it was not only "expedient" but also "proper" to set aside exclusive Native fishing grounds. The senior Fisheries official in British Columbia, A.C. Anderson, agreed. He wrote to Whitcher, shortly after O'Reilly's trip to the north coast, to stress the importance of setting aside land and associated fisheries for indigenous people. It was what he had done, as reserve commissioner, and it was what he understood O'Reilly to be doing as well.[15]

Even with this support for the reserve commissioner, Fisheries' refusal to recognize O'Reilly's exclusive fisheries prevailed. O'Reilly did not use the language of exclusive fisheries again, but access to fish remained central to many of his reserve allotments. This continued to concern Fisheries, and in response to O'Reilly's later work on the central coast (see Chapter 5), Commissioner of Fisheries Whitcher wrote that his department "[did] not recognize any unauthorized appropriations of public fishing rights by the Department of Indian Affairs for the exclusive use of Indians."[16] Indian Affairs then asked "that none of the Fisheries recommended to be set apart by the Indian Reserve Comm. for British Columbia ... be otherwise disposed of without the consent of this Dept."[17] In response, Fisheries restated its position:

> I have the honor to refer you again to the Minister's letters, stating that, as the common law and the statutes, now in force in those Provinces, entitle "every subject of Her Majesty" to use these fishing privileges, this Department cannot undertake to debar the public fishermen from exercising their legal rights in the premises, especially in view of the fact the ex parte reservations in question obviously exceed

the reason and justness of any arrangement founded in due regard for the relative rights of public fishermen and the necessitous claims of Indians.[18]

"Public fishermen" had "legal rights." "Indians," so far as Fisheries was concerned, only had "necessitous claims." This was an important distinction: rights outweighed claims, even "necessitous claims."

In what would be the final exchange in this intra-departmental debate, Indian Affairs indicated that, "in the event of any complications arising out of the diversion to other users of the Fisheries so recommended to be appropriated for the use of the Indians – [it would] hold the Dept. of Marine and Fisheries responsible."[19] There would be no short-term resolution; Indian Affairs and Fisheries simply disagreed over the respective rights of the "public" and Native peoples to the fisheries.

Whitcher's understanding of the law had emerged from his experience as an official in the fledgling Fisheries Branch within the Department of Crown Lands in the Province of Canada before Confederation (see Chapter 1). He was most familiar with the Great Lakes fisheries and with the conflicts there between Native and non-Native fishers. Between the 1830s and 1860s, a series of court decisions, none of which involved Native fisheries, suggested that the public right to fish was not limited, as in Britain, to tidal waters. Instead, it applied to the Great Lakes and other navigable waterways and thus restricted the Crown's prerogative to allocate exclusive fisheries in these waters.[20] The preponderance of legal opinion provided by government officials indicated that the public right to fish limited the granting of exclusive Native fisheries as well.[21]

This position has been taken up more recently by lawyer Roland Wright, who has argued that until the 1857 *Fisheries Act*, which provided that the Fisheries Branch could grant fishing leases of up to nine years, the Crown had no authority to allocate exclusive fisheries.[22] It was only with the consent of the legislature that exclusive fisheries, whether Native or non-Native, might be allocated in navigable waters, understood in North America as navigable in fact and including the Great Lakes and major rivers. Other scholars have vigorously challenged this argument on several grounds. First, this legal opinion did not accord with the reality of numerous instances in which the Crown had recognized exclusive Native fisheries in the eighteenth and nineteenth centuries, often through treaties.[23] Second, it misconstrued the source of the right to exclusive fisheries, placing it in Crown grants and not in the use and occupation of exclusive fisheries before the British assertion of sovereignty.[24] In this, Native claims were no different than those of settlers who had received a royal grant of an exclusive fishery from the French Crown before 1760 and thus before the British assertion of sovereignty.

Wright is correct, nonetheless, when he states that by the mid-nineteenth century, Fisheries officials believed that only they had the authority, under statute, to allocate exclusive fisheries. There is no doubt, moreover, that Whitcher believed this interpretation of the common law applied across Canada, including in British Columbia. As commissioner of Fisheries, he sent a circular outlining his views to all Fisheries overseers (local Fisheries officials) in Canada in 1875:

> Fisheries in all the public navigable waters of Canada belong prima facie to the public, and are administered by the Crown under Act of Parliament, which Statute imposes various restrictions on the public exercise of the right of fishing, and subjects the privilege to further regulation and control necessary to protect, to preserve, and to increase the fish which inhabit our waters.
>
> Indians enjoy no special liberty as regards either the places, times or methods of fishing. They are entitled only to the same freedom as White men, and are subject to precisely the same laws and regulations.[25]

Although there were no Fisheries overseers in British Columbia in 1875, and the first inspector of Fisheries for the province, A.C. Anderson, would not be appointed until the following year, it is clear from Whitcher's reaction to O'Reilly's work in 1881 that he believed this policy applied to navigable waters in Canada's westernmost province. In subsequent correspondence, discussed more fully in Chapter 5, Indian Affairs appears to have conceded that the reserve commissioner did not have final authority to allocate exclusive fisheries and that, just as the province and Dominion had to confirm the land grants, so Fisheries and Indian Affairs had to agree about the exclusive fisheries. Nonetheless, Indian Affairs fully expected that the reserve commissioner's recommendations would be implemented.[26] Moreover, Inspector of Fisheries Anderson and Indian Superintendent Powell supported O'Reilly's practice of allocating exclusive fisheries and thought it should continue. In this context, Indian Affairs argued that it would "appear therefore inexpedient, if not positively mischievous, to interfere with the system at present in vogue for allotting Fisheries to the Indian Bands of that Province."[27] However, Fisheries' position prevailed; the public right to fish would preclude confirmation of O'Reilly's exclusive Native fisheries.

The Public Right to Fish in British Columbia

The common-law doctrine of the public right to fish loomed over Peter O'Reilly's work as a reserve commissioner in the 1880s and 1890s. The Department of Fisheries refused to recognize exclusive Native fisheries and justified its refusal by reference to the public right to fish. As a result, O'Reilly, who had begun his work by allocating exclusive fisheries along important stretches of the Fraser and Nass rivers, stopped making such allocations.[28] For another eighteen years, however, he continued to allot reserves based on access to the fisheries, work that is considered in the following chapter. But although Fisheries' position with respect to exclusive fisheries was clear, it is much less clear that this position reflected the state of law in British Columbia, or even in Canada, in the 1880s. In fact, O'Reilly's allotment of exclusive fisheries in 1881, and his care to restrict them to non-tidal waters, closely matched contemporary developments in Canadian law.

The line of Ontario cases and legal opinions on which Fisheries relied, which suggested that the public right to fish applied to navigable, non-tidal waters, were decided or provided before British Columbia joined the Canadian confederation. In British Columbia, therefore, the court rulings had limited persuasive value and were certainly not binding. Moreover, in a series of British cases from the 1860s, including a House of Lords decision in 1863, the English courts had confirmed that the public right to fish applied only to tidal waters.[29] Unlike the cases from Ontario, the decisions of the English courts were binding in British Columbia. They might be distinguished based on different facts and perhaps even different geographical circumstances, but they could not be overturned. The scope of the public right to fish had not come before the courts in British Columbia, but it had arisen in New Brunswick and was the subject of an Exchequer Court of Canada decision in 1880, the year before O'Reilly began his field work.

In October 1880, Justice Gwynne of the Exchequer Court of Canada released his decision in *Canada v. Robertson,* a dispute over the right of fishing in the Miramichi River of New Brunswick.[30] Both the Dominion and the province claimed the right to allocate exclusive fisheries in the river, the Dominion under section 91(12) of the *British North America Act,* which placed "Seacoast and Inland Fisheries" within its jurisdiction, and the province under its ownership of and responsibility for property recognized in sections 109 and 92(13).[31] The case proceeded to Justice Gwynne as a series of questions about the rights of a riparian owner on a navigable but non-tidal river. Did the law create a presumption that a land grant next to such a body of water included the *solum* to the midpoint of the river – *ad medium filum aquae* – and thus the exclusive right of fishing in those waters?

Gwynne J. established that, whatever the law in Ontario, the right of fishing in New Brunswick rivers "must be considered with reference to the Common Law of *England,*" which restricted the public right to fish to tidal waters.[32] Although well aware of the Ontario cases, which he had surveyed in deciding an earlier case,[33] Gwynne J. did not consider them in *Robertson*. Instead he reviewed a collection of cases from the United States, some of which grappled with the dissonance of a common-law rule that treated navigable and tidal as synonymous in the North American context. In most of these cases, he concluded, the US courts held that while the public had a right of navigation in all navigable waters, the riparian owner held rights to the *solum* of navigable rivers and thus to exclusive fisheries above.[34]

From this analysis, Gwynne J. held that the public right to fish existed only in navigable rivers as understood in law to mean tidal rivers. Above the ebb and flow of the tide, rivers were "not open to the public for purposes of fishing, but may be owned by private persons, and in common presumption are owned by the proprietors of the adjacent land on either side, who, in right of ownership of the bed of the river, are exclusive owners of the fisheries therein opposite their respective lands on either side to the centre line of the river." Regarding the public right to fish and the relevance of Magna Carta, interpreted in the nineteenth century as the source of the public right to fish, he continued:

Magna Charta does not affect the right of the Crown, nor restrain it in the exercise of its prerogative of granting the bed and soil of any river above the ebb and flow of the tide, or of granting exclusive or partial rights of fishing therein as distinct from any title in the bed or soil, and in fact the Crown grants of land adjacent to rivers above the ebb and flow of the tide, notwithstanding that such rivers are of the first magnitude, are presumed to convey to the Grantee of such lands the bed or soil of the river, and so to convey the exclusive right of fishing therein to the middle thread of the river opposite to the land so granted.[35]

Even if the Crown had not granted the *solum* but had retained it for itself, Gwynne J. ruled that "such a reservation does not give to the public any common right of fishing in the river."[36] It remained open for the Crown, under its prerogative, to allocate exclusive fisheries in non-tidal rivers either as an incidence of a land grant or separately.

I have no evidence that O'Reilly had read or knew of the decision in *Robertson*, although he had been working as a lay judge before his appointment as reserve commissioner, but his allocation of exclusive fisheries in non-tidal rivers fit precisely within its parameters. As a representative of the Crown in right of both the province and Dominion, he could exercise the Crown's prerogative to allocate exclusive fisheries in non-tidal waters.

Officials in the Department of Fisheries, who must have known about *Robertson*, given that it was the minister of Marine and Fisheries who was being sued for licensing an exclusive fishery, showed no inclination to alter their position, even though the decision seemed to undermine their argument that the public right to fish prevented the Crown from allocating exclusive fisheries in non-tidal waters.

On appeal, in 1882, the Supreme Court of Canada upheld most of Justice Gwynne's decision. On the question of the public right to fish, the Supreme Court justices were less expansive. Chief Justice Ritchie wrote briefly that the portion of the Miramichi in question, which was navigable but non-tidal, "is not a public river on which the public have a right to fish," and although the public did have a right of navigation, "such a right is not in the slightest degree inconsistent with an exclusive right of fishing, or with the rights of the owners of property opposite their respective lands."[37] The public right to fish, according to the chief justice, did not apply in non-tidal waters. Justice Strong seemed inclined to disagree, suggesting that on "large navigable freshwater rivers, above the flow of tide,"[38] and on "the great lakes,"[39] the Crown might hold the right to fish in trust for the public. He was, however, reluctant to wade into these waters and concluded that the issue did not need to be resolved to decide the case.[40] Perhaps this hesitation of Justice Strong's was enough for officials at Fisheries. They maintained their opposition to O'Reilly's allocation of exclusive fisheries, even though, after *Robertson*, that position rested on highly unstable legal ground.

In 1896, fourteen years after the Supreme Court's decision in *Robertson*, Chief Justice Strong, as he then was, re-examined the application of the public right to fish in non-tidal waters. The case of *In Re Provincial Fisheries*, a Dominion/provincial reference to the Supreme Court over their relative powers to regulate fisheries, raised the rights of riparian owners and the extent of the public right to fish. On the public right to fish, there was no mistaking Chief Justice Strong's position:

> That the Crown in right of the provinces could grant either the beds of such non-tidal navigable waters or an exclusive right of fishing is, I think, clear. Before Magna Charta the Crown could grant to a private individual the soil in tidal waters with the fishery as an incident to it, or the exclusive right of fishing alone as distinct from the soil. Then, *as the restraint imposed by Magna Charta does not apply to any but tidal waters,* there is no reason why the prerogative of the Crown to make such grants in the class of waters now under consideration, large navigable lakes and non-tidal navigable rivers, should not be exercised now as freely as it could have been with reference to tidal waters before Magna Charta.[41]

In sum, the public right to fish did not restrict the Crown's right to grant land, including the *solum* of adjacent non-tidal lakes and rivers, to create exclusive fisheries. If the Crown held riparian land, or had reserved the fishery from a riparian land grant, and had not granted exclusive fisheries, then the public might fish. Indeed, Strong C.J. wrote that it had been the "invariable practice, which has prevailed in Canada from the earliest times since the settlement of the country, to treat the right of fishing in navigable waters above the flow of the tide as public."[42] As a result, the *ad medium filum aquae* presumption did not apply in navigable waters. Crown grants of land adjacent to navigable waters were not presumed to include the fishery to the midpoint. But this did not in any way compromise the Crown's prerogative to allocate exclusive fisheries where it saw fit.[43] This prerogative lay with the Crown in right of the province except in a few circumstances, such as Indian reserves, where it lay with the Crown in right of the Dominion.[44]

The Judicial Committee of the Privy Council heard an appeal of the *Provincial Fisheries Reference* but chose to defer the question of the public right to fish.[45] The Supreme Court's interpretation was neither rejected nor confirmed. Again, perhaps this equivocation was enough for the Department of Fisheries to continue to refuse to recognize exclusive Native fisheries in non-tidal waters. But it is hard to find a legal basis for that position. The preponderance of case law stood opposed. And after the 1913 decision of the Privy Council in *A.G. B.C. v. A.G. Canada,* there could be no doubt. Of the rights to fish within the Dominion-held railway belt (a forty-mile-wide tract of land that bounded the Canadian Pacific Railway through British Columbia), but also as a general statement about fishing rights, Viscount Haldane wrote:

> In the non-tidal waters they [fisheries] belong to the proprietor of the soil, i.e. the Dominion, unless and until they have been granted by it to some individual or corporation. In the tidal waters, whether on the foreshore or in creeks, estuaries, and tidal rivers, the public have the right to fish, and by reason of the provisions of Magna Charta no restriction can be put upon that right of the public by an exercise of the prerogative in the form of a grant or otherwise.[46]

The public right to fish restricted the Crown prerogative, whether the Dominion or the provincial Crown, in tidal waters only. The Crown was free to grant exclusive fisheries in non-tidal waters. To the extent that Canadian courts, including the Privy Council, considered the issue after British Columbia joined Confederation, this had been the consistent finding since Justice Gwynne's decision in *Robertson* in 1880, the year before O'Reilly began allocating exclusive fisheries in non-tidal waters.

Conclusion

If O'Reilly's allotments of exclusive fisheries are characterized as recognizing fishing rights that predate the British assertion of sovereignty, then the restrictions on what the Crown may grant are not an issue. The rights do not arise from Crown grant but precede the Crown and therefore are not subject to a common-law doctrine – the public right to fish – that limits the Crown's ability to alienate exclusive fisheries. Several authors have argued persuasively that this is the better interpretation, and it certainly accords with the recent direction of Aboriginal rights and title litigation. Aboriginal title and rights do not arise from Crown grant; nor do they depend on the application of the Royal Proclamation of 1763 to British Columbia. Instead, they arise from the long use and occupation of land and resources by Aboriginal peoples that predate British assertions of sovereignty.[47]

If, however, O'Reilly's allotments are understood as Crown grants that create exclusive fishing rights, then the common-law doctrine of the public right to fish might operate to limit what grants were possible. With respect to tidal waters, it is clear that the Crown could not alienate exclusive fisheries in tidal waters without authority from parliament. Beginning in 1857, with the first comprehensive fisheries legislation for the provinces of Canada (Quebec and Ontario), elected representatives provided the fledgling Fisheries Branch with the authority to allocate exclusive fisheries for up to nine years. After Confederation, this authority carried over in the *Fisheries Act* to the minister of the Department of Marine and Fisheries. Within these statutory limits, Fisheries could grant exclusive fisheries in tidal waters.

With respect to non-tidal waters, when O'Reilly began his work in 1881, he did so in the aftermath of a decision that clearly restricted the application of the public right to fish to tidal waters. In the ensuing years, this position would be confirmed repeatedly until, in 1913, a decision of the Privy Council left no doubt that the Crown's prerogative to grant exclusive fisheries in non-tidal waters was unaffected by the public right to fish. Opposition to O'Reilly's allocation of exclusive fisheries in non-tidal waters, based as it was on the doctrine of the public right to fish, was unfounded.

Nonetheless, the common-law doctrine of the public right to fish was a crucial part of the legal apparatus that surrounded Native peoples and dispossessed them of their fisheries. In tidal waters, it opened the fishery to everyone. In non-tidal waters, it provided the justification, albeit erroneously, to remove constraints on non-Native fishing. Native fishers could participate on the same terms as everyone else. However, this was the problem. In opening the fishery to everyone, the public right to fish (at least as interpreted by Fisheries) erased the prior rights of Native peoples to their fisheries. The radical simplification of Native customary tenure in the fisheries was achieved simply by denying its existence. A token Indian food fishery (discussed in Chapter 6) was all that would be allowed to remain. Beyond this erasure of tenure, the opening of the fisheries to the public undermined the Indian land policy that was premised on and justified by the continuing right of Native peoples to their fisheries. The land allotments only made sense in tandem with the fisheries. Finally, constructing the fisheries as common property, to which everyone had the right not to be excluded, worked best for those who had access to the credit and capital that was increasingly necessary to participate as fishers and, even more so, as processors. Native people shared a lack of access to capital with many immigrants, but their status as Indians under the *Indian Act* meant that credit was also difficult to acquire. In a legal regime that did not recognize their prior rights to the fisheries or the correlation between land policy and fisheries, Native peoples were never in a position to participate in the fisheries on equal terms.

5

Indian Reserves and Fisheries

The counting of people, the measuring of land, the scoring of trees, the planting of survey posts, the drawing of lines, and the registering of title: these were the steps in the construction of an Indian reserve geography in British Columbia, marking the land available to immigrants by carving out what was reserved for Indians. It was a formative delineation in a settler colony. The drawing of the boundary, logically prior to immigrant settlement, occurred in much of British Columbia after settlement had provoked unrest or conflict. As a result, the work of the reserve commissioners in much of the province was exceedingly difficult; the results left nobody satisfied. Certainly the vast majority of Native peoples thought the reserves provided an insufficient land base and that the best of their land had been lost. Conversely, many homesteaders begrudged the reserved land, believing that they or other immigrants could use it more productively. This was a nearly constant refrain from provincial politicians, many of whom pressed to have the reserves in their constituencies reduced. Despite these difficulties, and the inherent flaw of a process that did not address Native title, an Indian reserve geography did appear in British Columbia and remains largely intact to this day. No single person is more responsible for its implementation than Peter O'Reilly. Over his eighteen years as Indian reserve commissioner, including twenty-six circuits, O'Reilly allotted more than 650 reserves and confirmed nearly 200 allotted by his predecessor, Gilbert Malcolm Sproat.[1]

O'Reilly proved to be an efficient and economical commissioner and was in almost every respect a much safer choice than Sproat for a provincial government concerned with the size and scope of reserve allotments. Moreover, he kept things simple. Whereas Sproat sought numerous compromises, including commonages, temporary and provisional reserves, easements, and other informal arrangements, O'Reilly either allotted a reserve or not. However, as Chapter 4 revealed, he encountered rough water when he reserved fisheries. The Department of Marine and Fisheries (Fisheries) reacted swiftly and strongly to his setting aside of exclusive fisheries along the Fraser and Nass rivers in 1881. Not one to court controversy, O'Reilly stopped recording exclusive fisheries. Nonetheless, the fisheries remained central to the reserve geography that unfolded in the wake of his circuits. This chapter examines the changing ways in which O'Reilly recognized the importance of the fisheries to his reserve allotments. At the end, it looks briefly at the work of his successor, A.W. Vowell, who was able to accomplish very little, even though he was reserve commissioner for a dozen years, from 1898 to 1910. The province's hardening attitude towards reserved land effectively blocked its allotment until a jointly constituted royal commission began its work in 1913.

Reserves and Fisheries, 1882-98
Although O'Reilly stopped allocating exclusive fisheries after 1881, the instructions from the Department of Indian Affairs (Indian Affairs) had not altered: he was to "interfere as little as possible with ... fishing stations."[2] In June 1882, he travelled to Barkley Sound and the Alberni Canal, Nuu-chah-nulth territory on the west coast of Vancouver Island. During a week-long visit he allotted thirty-six reserves, mapped in Figure 5.1. All the reserves, which averaged 161 acres, bordered on the ocean or a river, and O'Reilly linked thirty-two of them, nearly 90 percent, directly to a fishery, including dogfish, halibut, herring, salmon, seal, and shellfish, or to materials needed in the processing and storing of these fish. Table 5.1 indicates the fishery or the fishing activity associated with each reserve. O'Reilly also told the Nuu-chah-nulth that their fisheries were protected. In response to enquiries from a Tseshaht chief, he replied that although alienated land could not be included within reserves, "the Government were anxious to secure to them all their fishing grounds."[3]

Two months later, in August 1882, O'Reilly moved into Kwakwaka'wakw territory, visiting the Gwa'sala-Nakawaxda'xw on the mainland opposite the northern tip of Vancouver Island. He told the Nakawaxda'xw, who were reluctant to show anyone their fishing sites for fear that outsiders would take over those places, that the purpose of his visit "was to secure the land to them, and prevent the possibility of either it, or their fisheries, being taken up by anyone."[4] He also recorded the following discussion with the Gwa'sala chief Neu-Kke-te, "who after the usual conversation, stated there was no farming land in his Country, but that he wished to have the site upon which their Village stood, and the fishery at the head of the Inlet secured to them, and to be assured that they would not be prevented from hunting on the mountain, or from fishing for halibut, and herring in

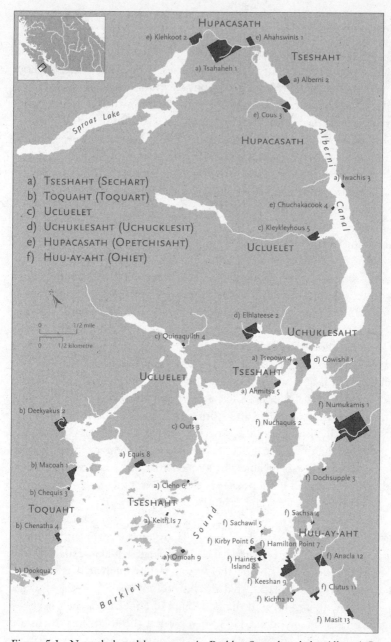

HUPACASATH
e) Klehkoot 2
e) Ahahswinis 1
a) Tsahaheh 1
TSESHAHT
a) Alberni 2
e) Cous 3
HUPACASATH
Sproat Lake
a) Iwachis 3
Alberni Canal
e) Chuchakacook 4
c) Kleykleyhous 5
UCLUELET

a) TSESHAHT (SECHART)
b) TOQUAHT (TOQUART)
c) UCLUELET
d) UCHUKLESAHT (UCHUCKLESIT)
e) HUPACASATH (OPETCHISAHT)
f) HUU-AY-AHT (OHIET)

d) Elhlateese 2
c) Quinaquilth 4
UCHUKLESAHT
a) Tseoowa 4
d) Cowishil 1
0   1/2 mile
0   1/2 kilometre
UCLUELET
TSESHAHT
a) Ahmitsa 5
f) Numukamis 1
b) Deekyakus 2
c) Outs 3
f) Nuchaquis 2
a) Equis 8
b) Macoah 1
f) Dochsupple 3
b) Chequis 3
a) Cleho 6
TOQUAHT
TSESHAHT
f) Sachsa 4
b) Chenatha 4
a) Keith Is 7
Sound
f) Sachawil 5
HUU-AY-AHT
f) Kirby Point 6
f) Hamilton Point 7
a) Omoah 9
f) Haines Island 8
f) Anacla 12
b) Dookqua 5
f) Keeshan 9
f) Clutus 11
Barkley
f) Kichha 10
f) Masit 13

Figure 5.1 Nuu-chah-nulth reserves in Barkley Sound and the Alberni Canal
allotted by Indian Reserve Commissioner O'Reilly, June 1882. Ninety percent
of these reserves were specifically connected to one or more fisheries, including
salmon, dogfish, halibut, herring, shellfish, and seal. (The spelling of the First
Nations' names as it appears in the historical record is in parentheses where it
differs from current usage.)

Table 5.1

Reserves and associated fisheries allotted by Peter O'Reilly in Barkley Sound and the Alberni Canal, June 1882

First Nation	Reserve	Fishery
Tseshaht (Sechart)	Tsahaheh 1	Salmon
	Alberni 2	Camping ground when returning from fishing
	Iwachis 3	Dogfish
	Tseoowa 4	Dogfish
	Ahmitsa 5	Dogfish
	Cleho 6	For fishing purposes
	Keith Island 7	Fishing station
	Equis 8	Shellfish, salmon, dogfish, seal
Toquaht (Toquart)	Macoah 1	Salmon
	Deekyakus 2	Salmon
	Chequis 3	Right of fishing in the creek
	Chenatha 4	Salmon
	Dookqua 5	Fishing station
Ucluelet	Outs 3	Fishing station
	Quinaquilth 4	Salmon
	Kleykleyhous 5	Salmon
Uchuklesaht (Uchucklesit)	Cowishil 1	Fish
	Elhlateese 2	Salmon
Hupacasath (Opetchisaht)	Klehkoot 2	Salmon
	Cous 3	Salmon
	Chuchakacook 4	Dogfish
Huu-ay-aht (Ohiet)	Numukamis 1	Dogfish, salmon
	Nuchaquis 2	Dogfish
	Dochsupple 3	Salmon
	Sachsa 4	Salmon
	Sachawil 5	Dogfish, salmon, herring
	Kirby Point 6	Fishing station
	Hamilton Point 7	Seal
	Kichha 10	Halibut
	Clutus 11	Halibut
	Anacla 12	Salmon
	Masit 13	Halibut

Source: Field Minutes, 6, 9, 11, 14, 16, and 17 October 1882, in Federal Collection of Minutes of Decision, Correspondence and Sketches, vol. 10 (P. O'Reilly [Indian Reserve Commissioner], June 1882 to February 1885). Copy held by Department of Indian Affairs, Vancouver Regional Office.

the sea." O'Reilly responded "that the mountains were as free for him to hunt upon as ever, and that he would enjoy the right to fish in the Ocean *in common* with others."[5] This characterization of a right to fish "in common" emulated the fisheries clause of the 1854-56 treaties in the Washington Territory.[6] Perhaps this was inadvertent – O'Reilly offered no further written explanation for his choice of language, and he does not use the phrase again – but it was another indication of the centrality of the fisheries in his reserve allotments.

Later in August, O'Reilly arrived in Heiltsuk territory on the central coast. His work there provides another good example of what he was doing along the coast. Of the twelve reserves allotted to the Bella Bella (O'Reilly divided the Heiltsuk into Bella Bella and Kokyet), the first secured their

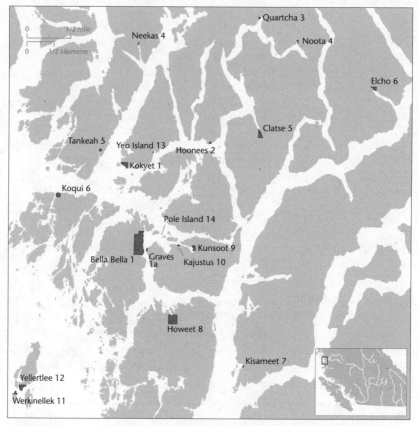

Figure 5.2   Heiltsuk reserves allotted by O'Reilly, August 1882. He identified all but three reserves, which included a grave site and the principal village site, as fishing stations.

principal village site and the others were intended to secure access to salmon, halibut, and seal fisheries (Figure 5.2). Using language that he would repeat many times for coastal reserves, O'Reilly described Bella Bella Reserve No. 10, a fifteen-acre parcel, as "only a fishing station, and of no value for any other purpose." Reserve No. 11, containing sixty acres, was the principal summer residence, and O'Reilly wrote of it: "The Western shore is rugged, and weather beaten, and the Reserve of no value except as a fishing station."[7] Of the six reserves he allotted to the Kokyet, four were identified as fishing stations.

O'Reilly no longer mentioned exclusive fisheries in his reserve allotments, but otherwise he continued much as before, setting aside small parcels of land to secure access to fish, the only resource of any consequence that might provide the Nuu-chah-nulth, Kwakwaka'wakw, and Heiltsuk peoples with a reasonable livelihood. Native peoples seem to have been well aware of this pattern, and they sought to secure their fisheries in ways that fit within it. A Haida petition for additional reserves in 1883, for example, requested eight single-acre plots that would secure access to, and enable processing of, dogfish, salmon, and halibut.[8] O'Reilly's reports confirm that this was a sound strategy. In 1886, he informed the Kwakwaka'wakw chiefs of the Lower Knight Inlet tribes "that the object of [his] visit, was to secure to them certain plots of land which would give them the control of their fisheries."[9] In his attempt to convince the Lekwiltok tribes to cooperate in identifying land they wished him to reserve, O'Reilly was a little more tentative; he claimed to have "pointed out to them the advantages they would derive from having lands so set apart, which would virtually give them the control of their fisheries."[10]

O'Reilly's language was not consistent, but it is unlikely that these assurances, however framed, provided much comfort to Native peoples who were witnessing the increasing penetration of immigrant settlers and a state-based legal system into their daily lives. Indian Affairs was also concerned that although O'Reilly appeared to be making general guarantees about Native access to fisheries, they were not specific enough. After reviewing the records from O'Reilly's 1888 tour through Sliammon, Klahoose, and Homalco territory on the coast north of Vancouver, one official noted that although there was much mention of fishing stations and important river fisheries, nowhere did O'Reilly specify the waters that were to be set apart for exclusive Native use. O'Reilly responded that all the fisheries he was being asked about were in tidal waters and he was "not aware that it has ever been the practice to assign exclusive fishing rights in these." Further, he was not aware of any authority for doing so.[11] In this aspect, at least, O'Reilly had been consistent.

In 1889, O'Reilly spent most of June in Nuu-chah-nulth territory on the west coast of Vancouver Island (see Appendix, Figure A1). At Tofino Inlet and around Kennedy Lake, he allotted ten reserves, all of which he specifically linked to the fisheries, principally salmon. Two of these reserves, Clayoqua 6 and Winche 7, were located on rivers flowing into Kennedy Lake, and O'Reilly stipulated that the Clayoquot had the "right to fish" in the rivers running through the reserves. Another reserve, Ilthpaya 8, at the outflow of Kennedy Lake, included the "right to fish" in the Kennedy River from the boundary of the reserve downstream a distance of approximately one mile to the tidal waters. The following year the provincial government sold 160 acres of land surrounding this fishery to a firm that intended to build a salmon cannery on the site. Did the Clayoquot "right to fish" in the Kennedy River place any restriction on the cannery operations? Thomas Mowat, appointed inspector of Fisheries in 1886, thought it should not. It was, he argued, "unwise to grant the Indians an exclusive right to fish in any of the above named streams for as the Country begins to be settled it is likely to make trouble."[12] Moreover, he thought the 1888 fisheries regulations, which provided that "Indians shall, at all times, have liberty to fish for the purpose of providing food for themselves," gave all the protection that was required.[13] O'Reilly himself denied that he had allocated an exclusive fishery. Rather, the Clayoquot could not be prevented from taking fish, but nor could anyone else; they were common property. On this understanding, reserve grants that were intended to secure access to the fisheries, in fact provided nothing more than the public's right not to be excluded.

The issue arose elsewhere in the province as well. In 1890, a dispute flared at Lowe Inlet over the presence of cannery boats in waters that the Kitkatla believed O'Reilly had reserved to them in 1882 (Figure 5.3). Chief Shukes told the local Fisheries Guardian: "Judge O'Reilly gave this land and water to my people, I do not want any Whitemen to fish here please tell your chief I have fished at Low's Inlet for 8 years. It is the principal support of myself and people."[14] This particular dispute was resolved when the cannery agreed to purchase its fish from Kitkatla fishers. Four years later, however, Fisheries officers charged the Lowe Inlet Canning Company for purchasing fish from unlicensed Native fishers.[15] In 1905, Indian Affairs again expressed concern that the licensing of non-Native commercial fishers in tidal waters south of the Skeena River, including Kitkatla territory, would damage the fisheries that the reserves had been intended to protect.[16]

Seeking some clarity on the authority of the reserve commissioner, Fisheries asked the Department of Justice for a legal opinion. In this, Deputy

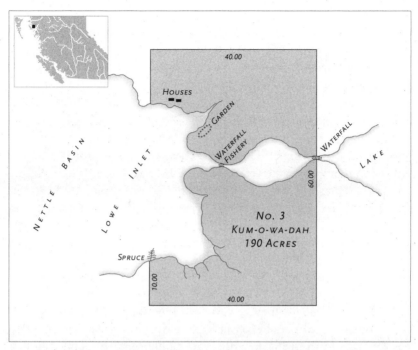

Figure 5.3 O'Reilly's sketch of the Kitkatla Reserve No. 3 Kumowdah, allotted 21 September 1882.

Minister Robert Sedgewick appeared to confirm Fisheries' position: "The Indian Reserve Commissioner has not the power to set apart for the exclusive use of the Indians any of the waters of British Columbia."[17] But, he continued, it lay within Fisheries' jurisdiction to act on the commissioner's recommendations to set aside exclusive fisheries. Lawrence Vankoughnet, deputy minister of Indian Affairs, replied that his department had never claimed that jurisdiction for the reserve commissioner; that just as the reserve lands had to be approved by the provincial and Dominion governments, so the fishing rights had to be approved by Fisheries. However, he fully expected that Fisheries would not dismiss them out of hand, as it had been doing. "It appears," he argued, "only reasonable and just that Indians owning Reserves should have the fisheries opposite or close to those Reserves excluded from license to other parties."[18]

Charles Tupper, minister of Fisheries, was not impressed with these arguments. The fisheries regulations provided all the protection Native fishers required, he contended, and he asked Indian Affairs to instruct O'Reilly and the Indian agents to convey to their charges that their food fishery was a valuable privilege, conferred "as an act of grace" upon them by a benevolent

Dominion government, and one that would be withdrawn if abused.[19] Edgar Dewdney, minister of the Interior and responsible for Indian Affairs, capitulated. Instructions were sent to O'Reilly and the Indian agents in British Columbia to inform Natives that their access to the fishery was a privilege that the Dominion government might revoke.[20]

After this exchange, O'Reilly again modified his treatment of the Native fisheries. Instead of attempting to provide additional protection for Native fisheries by reserving the land at important fishing sites and stipulating rights to fish, he claimed that the fisheries regulations provided all the protection required. In a meeting with the Lake Babine at Wit-at Village, site of the Hudson's Bay Company's (HBC) Fort Babine, O'Reilly responded to a request to reserve an old weir-fishing site by stating: "I do not consider it necessary to make a fishery down the river, for the law forbids a white man to use a net in fresh water so that all the salmon is for the Indian."[21] Several weeks later he told the Gitxsan at Gitsegukla that he "would include their principal fisheries, in the immediate neighbourhood, but there is a special Act which prohibits white men from fishing (in fresh water) therefore the fisheries are virtually all the Indians, and it is not necessary to make a reserve of every little place where an Indian fishes."[22] He was referring to the 1888 fisheries regulations, which prohibited fishing for salmon with nets in fresh waters, except for the Indian food fishery.[23]

In 1897, when O'Reilly's work as Indian reserve commissioner was completed, Indian Affairs asked him about his fisheries allotments. He replied that he had been "governed to some extent" by the approach taken by his predecessors – the Joint Indian Reserve Commission and then Sproat – and also by his directions, given seventeen years earlier, that he was to interfere as little as possible with the Native fisheries and that the fishing stations he allotted were to be defined clearly. He understood that copies of his reserve descriptions had been forwarded to the Department of Fisheries, but he claimed not to have corresponded directly with the department or to know its views. This must have been feigned ignorance, given the debate that had swirled around his allocation of Native fisheries throughout his tenure as reserve commissioner. O'Reilly concluded, emphasizing the importance of the fisheries to the whole process:

> The various Commissioners, as also the Indian Agents, have impressed upon the Indians that reserves so made and approved could not be trespassed upon; and that their rights and privileges would be guarded: It would, therefore, be a hardship, & a source of regret, if by any act of the Marine & Fisheries Department, faith should be broken with them.

In many instances where it has been found impossible to provide agricultural lands, the Indians rely almost entirely on their fisheries for support.[24]

Indian Affairs then set out to catalogue whatever fishing rights or privileges had been accorded Native peoples across the country. Every Indian agent was asked to outline "what fisheries, if any, have been reserved for your Indians, describing them and the privileges they now enjoy; whether they observe the fishery regulations – and especially those regarding the close season; what complaints, if any, the Indians have in regard to fisheries, and what special privileges, if any, you consider should be accorded your Indians."[25]

According to the Indian agent for the Nuu-chah-nulth, most of whose reserves O'Reilly had allotted in two visits in 1882 and 1889, approximately 60 of their 150 reserves were adjacent to or surrounding salmon rivers, and "it was understood by the Indians that they had fishing rights in all these streams, some of the reserves being given for that reason only." The agent for the Nisga'a and Tsimshian reported similarly that they "enjoy the exclusive right to fish in the fresh water streams where fishing reserves are located."[26] The Nlha7kapmx and Okanagan agent identified particular fishing stations. Agents for the Kwakwaka'wakw and Coast Salish indicated that the first section of the fisheries regulations provided the primary protection for Native fisheries in their jurisdictions. Some of the Indian agents reported that little more was needed in the way of additional protection for the Native fisheries, particularly in the regions where there was little interest in fishing among the immigrant population (including the Kootenay, Williams Lake, and Babine agencies) and hence limited competition for the resource. Overall, however, the response from the Indian agents revealed the tenuous position of Native fisheries in British Columbia.

Using the information from the Indian agents, Indian Affairs produced a list of all the exclusive fisheries that had been set aside and all reserves that had been allotted as fishing stations in British Columbia. The list amounted to nearly 600 locations, the vast majority of which were identified as fishing stations.[27] Slightly more than 1,000 reserves had been allotted by 1898,[28] so by the reckoning of officials in Indian Affairs, nearly 60 percent of these reserves were intended to ensure Native peoples' access to their fisheries. In addition, the fourteen bands on Vancouver Island that were parties to the Douglas Treaties had their right to fish protected by those treaties.

Indian Affairs also began to outline the legal argument that might be used to support its position. It went something like this: When British Columbia joined Confederation, jurisdiction over "Indians, and lands reserved for Indians," was transferred to the Dominion government. Anticipating this transition, the thirteenth article in the Terms of Union between British Columbia and Canada required the Dominion to pursue as liberal a policy towards Natives as had the colonial government. The former colony had paid little attention to the fisheries, but the best evidence of colonial policy towards the Native fisheries was contained in the Douglas Treaties, which secured the Natives a right to their "fisheries as formerly." The Dominion was bound, therefore, to protect the Native fisheries. Moreover, the agreement between the two levels of government establishing the Joint Indian Reserve Commission, and the instructions to the reserve commissioners, directed them not to disturb Native fisheries and to set aside fishing stations. Finally, the opinions of A.C. Anderson as the inspector of Fisheries and Gilbert Malcolm Sproat as reserve commissioner, frequent defenders of the Native fisheries, were proffered to bolster the argument.[29]

But the conclusion of all this work was disappointing to those who sought greater protection for the Native fisheries. The Department of Fisheries claimed jurisdiction to recognize exclusive fisheries, whether for Native or non-Native fishers, and in the case of the former, it simply refused to do so.[30] At the end of another long memorandum, which incorporated much of the evidence and many of the arguments outlined above, J.D. McLean, secretary Indian Affairs, concluded: "As regards British Columbia, the undersigned does not see what further can be done at present than to rely on the good offices of the Minister of Marine & Fisheries to deal fairly with any grievances of the Indians that may be brought to his knowledge."[31] Given Fisheries' record, this was scant consolation.

Vowell and the Indian Reserve Commission, 1898-1910

Peter O'Reilly's tenure as reserve commissioner drew to a close not only as the Department of Fisheries hardened its position against Aboriginal fisheries and Indian Affairs conceded its impotence, but also as the provincial government's opposition to further reserve allotments effectively brought the process to a standstill. In comparison to the 650 reserves that O'Reilly allotted and the 200 more of Sproat's that he accepted,[32] the provincial government was prepared to confirm just 49 reserves between 1898, when A.W. Vowell assumed the role of Indian reserve commissioner, and 1910. Ten other reserves were added during Vowell's tenure through purchase, transfer, or donation, but the scale of reserve creation was small compared to that during O'Reilly's term. Dominion/provincial relations were poor, the question of Indian reserves as well as jurisdiction over fisheries were among the irritants, and much of Vowell's work was ignored by the province, which was increasingly unwilling to allot more land to Native peoples and concerned about what had already been set aside.[33]

Vowell had been the Indian superintendent in British Columbia since 1890. He became Indian reserve commissioner in 1898 and continued in both roles until 1910. He was well aware of O'Reilly's work and of the important connections between reserved land and fisheries. Of Vowell's forty-nine confirmed reserves, he specifically mentioned fisheries as a reason for allotting twenty-one of them; the report of the royal commission in 1916 identified another six of these as fishing stations. Moreover, Vowell understood the reserves to include the fisheries. Of two Sechelt reserves, Sekaleton 20 and Saughanaught 22, he wrote: "These allotments, which include some fisheries, houses and gardens, were assigned to them."[34] Similarly, for the two Lhoosk'uz Dene' (Kluskus) and Ulkatcho allotments, Vowell wrote that the "reserves, which include their fisheries, hay meadows and gardens, have been laid off for them."[35]

One set of reserves allotted during Vowell's tenure arose in unique circumstances. In 1906, the Lake Babine and the Department of Marine and Fisheries settled what became known as the Barricade Treaty. The Babine agreed to remove their fish weirs from the Babine River, a tributary of the Skeena and an important sockeye spawning ground, in exchange for nets, a school, agricultural supplies, and additional reserve land. Given the virtual stalemate between the two levels of government over reserve allotments, the Dominion purchased land from the province to honour its commitment for additional reserve land under the treaty. Fisheries provided the nets; Indian Affairs provided the funds to acquire nearly 2,000 acres of land, spread over four reserves. The largest of these new reserves, Babine IR 16, was on the west bank of the Babine River, opposite the principal Babine village, the Oblate church, and the HBC post, at the site of an age-old weir fishery. The other reserves were all on the shores of Babine Lake.[36]

By the early twentieth century, however, the work of the reserve commissioners who had allotted land in order to secure access to fish seemed of little consequence. Native peoples watched from their tiny land base as their once valuable fisheries, which might have supported viable local economies, were reallocated to other users. Native fisheries had been detached from reserved land.

Conclusion
After the ruckus over his "exclusive fisheries" in 1881, Peter O'Reilly stopped using that language but continued to reserve land to secure access to local

fisheries. He recognized that, in most cases, fish were the only resource that might have made reserve economies viable. But although fish became no less important, O'Reilly's ability and perhaps even willingness to secure Native access through the reserve commission diminished. Having begun with exclusive fisheries, O'Reilly, by the 1890s, pointed to the food fishing provisions in the fisheries regulations as sufficient protection, in some cases declining to grant fishing reserves because he felt they were no longer needed to secure access to the fish. He bent with the prevailing winds that were blowing from Fisheries, and so, for the most part, did Indian Affairs.

Although officials within Indian Affairs spoke out against the policies of the Department of Fisheries from time to time, there was never a sustained institutional commitment to secure broad recognition of Native fishing rights. Indian Affairs represented a marginalized community without much voice in Ottawa, and it was part of a series of Dominion governments that conceded the question of Native title in British Columbia to a province that denied its existence. Indian Affairs also conceded the prior existence of Native fishing rights to the Department of Fisheries. The latter had a powerful constituency in the cannery operators, who demanded access to the resource. Native rights presented an obstacle to the cannery interests, but it was an obstacle that could be and was largely ignored in the late nineteenth and early twentieth centuries. Yet even as Indian Affairs deferred to the more powerful department, the questions of Native title and fishing rights remained a shadowy presence. The following remarkable passage, written in 1913 by Thomas Deasy, the Indian agent for the Queen Charlotte Agency, reflects the foundational ambivalence within Indian Affairs to Native title and fishing rights:

> In conclusion, permit me to state that we are dealing with Indians who have supposed grievances on land and other questions. They come from a class that ruled the whole country less than a century ago. Where we find other people infringing the law and taking up the old Indian settlements not included in the reserves apportioned by former commissioners, the Indians lay before us papers in which they are told that they have certain rights to rivers and hunting grounds. When they go there, and find the hunting lodges torn down, and the ground occupied, they are discontented. It was a mistake, in the early days, for people, presumably of authority, to give the Indians these papers. At the present time, when the canneries want fish, and the Indians desire to use the streams, they find that others are poaching, and the chances of supply giving out are apparent. Our Indians will not take to the land until the fish disappear in our waters. Many complain of the Indians not making use of their reserves; but we must remember that they are bringing food from

the sea for the large local and export trade. They cannot fish and farm at the same time, and it is questionable whether these people could produce more wealth from the soil than they can from the water. So far as farmers are concerned on these islands, it means years of toil for any person taking up land even to clear off the timber. It is not farming country, and the surrounding waters are considered the best for fish in the world. The Indians are sought for by the owners of the cold-storage and cannery plants, and hundreds of fishermen in any country produce wealth ... It has been remarked that British Columbia headed the list in the production of fish during the past year, and to the Indians must be ascribed a great deal of the credit for the wealth produced in the surrounding waters. When the time comes for the Indians to take to the land, the Haidas will be as ready to take their place there as any other race. At present they are needed to garner the sea, on the shores of which they have always spent their lives. There are no better fishermen, or boat-men, in any country, and the pity of it is that they cannot gain all they make from the water, in the way of marketing their catches.[37]

Ambivalence towards rights and title pervades this passage, as it did Indian Affairs. The Haida grievances are "supposed," and yet Deasy recognizes that they stem from an era when the Haida were the undisputed rulers of the islands and surrounding waters. It would have been better had their rights never been recognized in any form, he suggests, and yet they have been. The Haida have papers, perhaps from the Indian reserve commissioners, recognizing at least some of their claims to hunt and fish, and Deasy is clearly sympathetic to the Haida efforts to retain those rights. He also exhibits considerable admiration for their prowess as fishers, but there is a sense that they will not become equals among civilized peoples until they "take to the land," an inevitable occurrence, Deasy believes, for which the Haida are well placed. In the meantime, they ought to be allowed to sell their catches.

Occasionally the ambivalence becomes certainty, and there are moments when Indian Affairs stood its ground against Fisheries. The 1925 case of *R. v. Charlie,* described in the following chapter, is one example. Indian Affairs opposed Fisheries in court over the closure of a Squamish food fishery on the Capilano River. But the moment was short-lived. Fisheries' position prevailed on appeal, and Indian Affairs did not pursue the matter. In truth, it was more interested in limiting the costs it would incur to support Native peoples, who had become destitute from their lack of access to resources, than in establishing fishing rights.[38] With a weak advocate in government and no access to the courts themselves, Native peoples had few tools with which to oppose their loss of control of the fisheries.

6
Constructing an Indian Food Fishery

In the last quarter of the nineteenth century and the first of the twentieth, while the reserve commissioners inscribed the legal geography of Indian reserves on British Columbia, the Dominion of Canada established the basic tenets of its fisheries regulation on the Pacific coast. In doing so, it constructed an Indian food fishery in law, providing what amounted to limited and discretionary protection for subsistence fishing. This food fishery mirrored the emerging reserve system; they arose together and were the means by which the Dominion opened the fisheries and the province opened the land to newcomers.

In this chapter I review the legislative and regulatory framework that surrounded and, in large measure, constructed the Indian food fishery. The analysis builds upon earlier work in which I describe the imposed legal structures and the reactions to them in greater detail.[1] Here I extend that analysis into the twentieth century, tracing the continuing effort to confine Native fishing to subsistence fishing and linking it to the creation of Indian reserves. The goal of senior Fisheries officials throughout was to maximize the fish available to other users, primarily the canning industry, but also to sport fishers as the case of *R. v. Charlie* reveals. Protection for Native fisheries would extend only far enough to prevent Native peoples from becoming a financial burden on the state, although even that modest goal was all too often jettisoned in order to maximize fish for the canneries and sport fishers.

Constructing an Indian Food Fishery
The colony of British Columbia joined the Canadian confederation in 1871, ceding authority over fisheries to the Dominion government. The Dominion *Fisheries Act* of 1868, which became law in British Columbia in 1877, surrounded the fisheries with a new regulatory framework. It empowered

the Crown to appoint Fisheries officers, authorized the minister of Fisheries to issue fishing leases or licences up to nine years in length, established some basic rules about how and when fish might be caught, set out fines for the failure to comply, designated Fisheries officers as justices of the peace with powers to convict and punish for offences, and gave the minister authority to make regulations for the better management of the fishery.[2] Indians were mentioned in the act only to indicate that the Department of Fisheries might allow them, at its discretion, "*to fish for their own use.*"[3] This provision, which gave Fisheries complete discretionary authority over Native food fisheries, had its origins in the management of Great Lakes fisheries.

Historian Michael Thoms has demonstrated a correlation between the early-nineteenth-century treaties in Upper Canada and subsequent legislation that restricted settler access to the fisheries. The statutes exempted Native peoples from the fisheries regulations and limited non-Native access. The Crown, Thoms argues, was following through on its treaty commitments to protect established Native fisheries.[4] But with growing non-Native interest in the fisheries, the pressure to open the protected fishing grounds increased. In 1857, the Province of Canada passed the first comprehensive fisheries legislation, introducing substantial amendments the following year.[5] There was no mention of or protection for Native fisheries in either act, something Thoms attributes to the growing political influence of sport fishers, an elite group who sought to preserve the fisheries for their enjoyment. Non-Native commercial fishers also gained better access through provisions that allowed long-term leases of commercial fishing stations. Native fishers were nominally eligible to acquire leases, but this seldom occurred. Instead, commercial fishers received leases to favoured Ojibwa fishing grounds on Lake Huron, provoking extended conflict.[6] As a result, when British Columbia joined the Canadian confederation and ceded jurisdiction over fisheries to the Dominion government, it acquired a set of fisheries laws and a bureaucracy of Fisheries officials that were the products of an extended and continuing colonial encounter.

In 1878, there were eight salmon canneries and three salting operations on the Fraser, two canneries on the Skeena, and salting operations on the Nass River and in Alert Bay. Approximately 1,200 fishers working with gill nets from nearly 400 boats brought salmon for processing to 1,600 shoreworkers.[7] Responding to the need to regulate this burgeoning industry, the Department of Fisheries established the *Salmon Fishery Regulations for the Province of British Columbia*. This first set of regulations consisted of three sentences that confined the fishery to tidal waters, restricted net sizes, and closed the fishery on weekends:

1. Drifting with salmon nets shall be confined to tidal waters; and no Salmon net of any kind shall be used for Salmon in fresh waters.

2. Drift nets for Salmon shall not be so fished as to obstruct more than one-third of the width of any river.

Fishing for salmon shall be discontinued from eight o'clock A.M. on Saturdays to midnight on Sundays.[8]

If they had been enforced, these regulations, which banned nets in fresh waters without an exemption for Native fisheries, would have closed many important, predominantly net-based salmon fisheries in the rivers of the province's interior (Photo 6.1). They provoked a sharp reaction from the reserve commissioner at the time, Gilbert Malcolm Sproat. What, he asked,

Photo 6.1   Native dip-net fishery on the Fraser River, ca. 1890. The fisher working his net is standing on a wooden platform over the river. Note the second, somewhat higher, platform just downstream. It might have been used a few days or weeks earlier when the river would have been higher. The sockeye salmon displayed on the plank might be the product of several minutes or several hours of work, depending on the size of the run. | *BC Archives A-06077*

was he to do when the Department of Fisheries had, in effect, closed the Native fishery in the interior of the province?[9] Fisheries did not change the regulations but responded instead with an informal exemption for Native fishers. Inspector of Fisheries A.C. Anderson, formerly of the Joint Indian Reserve Commission, was instructed "to suspend the application in regard to Indians" and "practically to exempt Indians from the operation of the Fisheries Regulations."[10]

As the canning industry grew over the decade following the 1878 regulations, and as the number of Fisheries officials in British Columbia increased, Native fisheries became the target of increasing supervision and control. At first the department turned its attention to those who participated in the industrial commercial fishery on the Fraser without a licence, but it soon concerned itself with Native fisheries on the rivers that drew anglers, particularly along the east coast of Vancouver Island. The informal policy not to enforce the *Fisheries Act* and *Regulations* against Native fishers became instead a focused effort to limit Native fishing as departmental personnel changed and as non-Native interest in sport and commercial fisheries grew. Anderson, who had sought to temper attitudes towards the Native fisheries, died in 1884 and was replaced by inspectors who sought to manage the resource by controlling the Native fisheries.

In 1888, Fisheries rewrote the regulations for British Columbia, making three significant additions: fishers were required to register their equipment and intended fishing location; Fisheries could limit the number of boats in a region; and, under certain circumstances, Indians were not required to hold a licence. This final clause read as follows: "Provided always that Indians shall, at all times, have liberty to fish for the purpose of providing food for themselves but not for sale, barter or traffic, by any means other than with drift nets, or spearing."[11] The wording was ambiguous, but the intention was to exempt most forms of Indian food fishing from licensing requirements and restrictions. All commercial fishing, however, had to occur under Dominion licence. A similar provision was included in the single clause pertaining to trout fisheries.[12]

These provisions allowed Native peoples virtually unfettered access to a food fishery, while at the same time placing no formal restrictions on their participation in the commercial fishery. What the provisions did not do was provide any protection for the Native fisheries. Commercial drift-net licences were available to anyone to fish for salmon in tidal waters, and canneries were allowed seine licences. These licence holders increasingly occupied important Native fishing grounds, provoking concern in many Native communities about access to the fisheries and about the viability of the harvest.[13] Native peoples attributed declining runs and food fish harvests to the cannery boats operating at the mouths of rivers along the coast. The regulations creating the food fishery, and even the reserves that had been allotted to secure access to the fisheries, provided no protections against these depredations.

Whatever limited protection the food fishery provided in law, it offered even less in practice. In 1892, Fisheries established a commission under the direction of Samuel Wilmot, Dominion superintendent of Fish Culture, to report on the state of the fishery in British Columbia and to make recommendations for its regulation. He had been asked to write a quick report in 1890 and had done so, but in 1892, he and the other commissioners heard from 112 witnesses, primarily in New Westminster, but also in Victoria, Nanaimo, and Vancouver. It appears that only two Natives testified: Charlie Caplin of the Musqueam and Captain George of Chehalis. George indicated that numerous others from his community had wanted to testify, but that the Indian agent, Mr. Tiernan, who also acted as the interpreter, had indicated that one was enough. Both Caplin and George complained that since Fisheries introduced the limited-licence regime on the Fraser in 1889 (a short-lived experiment discussed in Chapter 7), Native fishers had not had sufficient access to commercial licences, which they said were going to whites. George testified that Fisheries officers were also restricting their food fishery. To his claim that "God gave them [the Natives] these waters and the fish and the land, and now it is taken away from them by new comers," Wilmot replied, through Tiernan: "You tell him that the law gives preference to them – that they can fish without licenses for their own use, but not for barter or sale, and that when they come in competition with white men, they must stand in the same position as white men, but when fishing for their own use, they can fish without licenses." Indian Agent Tiernan responded: "But I may tell you, Mr. Wilmot, that they are not allowed to fish. I know an instance where their nets were taken and cut to pieces up at Yale – a poor cripple of a man – and they have not the privileges you speak of."[14]

In 1894, following the recommendation of the 1892 commission, Fisheries extended the *Regulations* to thirty sections. These revisions provided local officials with the first detailed set of rules to govern the fishery in British Columbia. They also allowed for greater state intervention in the Native fishery. An Indian food fishery remained exempt from the licensing requirements, but access now required a permit: "Provided always that Indians may, at any time, *with the permission of the inspector of fisheries,* catch fish for the purpose of providing food for themselves and their families but for no other purpose." In another section containing a general prohibition of nets in fresh water, there was a similar exemption: "But Indians may, *with the permission of the inspector of fisheries,* use dip-nets for the purpose of providing food for themselves and their families, but for no other purpose."[15] The Indian food fishery was not to be considered a right, wrote the minister of Fisheries, Charles Tupper, but "an act of grace" bestowed by a munificent state.[16]

The requirement that Native fishers acquire a permit to participate in the food fishery symbolized and, finally, effected the legal capture of the fisheries.[17] The state had not yet allocated sufficient resources and personnel to enforce the requirements, but the legal structure of capture was in place. Somewhat counter-intuitively, the state had captured the resource with a legal regime that, with few restrictions, opened the fisheries to everyone. Apart from allowing Native peoples a limited food fishery, the Department of Fisheries ignored the question of prior Native ownership.

Minimizing the Indian Food Fishery

In managing the food fishery, Fisheries' goal was to contain what it understood to be a privilege. After 1894, Native fishers needed a permit from the inspector of Fisheries to conduct their food fisheries, but it is not clear when that requirement was first enforced. Fisheries amended the regulations in 1908, following the recommendations of the Dominion-British Columbia Fisheries Commission 1905-7 to remove the need for a permit, instead requiring that "Indians and explorers in unorganized districts" report their food fishery catches. In 1910, the department restored the permit requirement, and a system evolved whereby the inspector would issue food fishing licences to Native fishers on the advice of the local Indian agent.[18] At times the inspector would challenge the list and the discretionary power of the Indian agent. Those Natives who worked or who were deemed capable of working in the wage economy were not eligible for food fishing permits, which were to be restricted to those who needed to fish to support a bare subsistence. At other times the inspector would go further, closing a river or other region to food fishing, while allowing commercial or sport fisheries.

The limited food fishery did not, of course, stop Natives from selling or trading fish without a licence. Fisheries was most concerned about Native fishers who caught and sold fish to canneries without a commercial licence. What success it had in limiting these sales is uncertain, but the fact that the fishery was illegal, and therefore posed an additional risk to purchasers, reduced the price that Native fishers working in this informal economy received for their fish. Native fishers also continued to supply fresh fish to local markets. The efforts of local Fisheries officers to control these fisheries varied considerably. Enforcement sometimes involved formal charges under the *Fisheries Act,* appearances before local magistrates, and the levying of a fine. The success of these formal proceedings was mixed. Convictions were elusive, often because of sympathetic magistrates who reflected local sentiment that Native fishers should be left alone. James Teit, ethnographer of the Interior Salish and advisor to the Allied Tribes (an organization of coast and interior First Nations formed to advocate for recognition of Native rights and title), thought the reluctance to confine the Native fishery was widespread: "The sentiment of all classes of white persons in British Columbia, including magistrates, police, fishery Officers, Indian Agents, lawyers, merchants, traders and ranchers is entirely in sympathy with the Indians in the matter of this prohibition, and opposed to any law which would prevent the Indians from obtaining their natural and necessary food."[19] Although the position is probably overstated, acquittals were common. In these circumstances, Fisheries officials relied on other techniques to contain the food fishery, notably surveillance, confiscation of fishing gear, and harassment. This occurred under the authority of law but beyond the purview of the local legal forum of the provincial court.

There was considerable support for the idea that Native fisheries, including fish caught for food and for sale in local markets, should be protected. Some proposed a "peddler's licence" that would formally sanction the widespread practice of catching fish in small quantities to sell.[20] But the Department of Fisheries was not interested in these proposals. Its officials sought to ensure that fish caught under food fishing permits were for the consumption of the community and not for sale, even if the sale were to raise money to purchase other foodstuffs. In 1913, a stipendiary magistrate at New Westminster alerted Fisheries to a potential problem with its definition of Indian food fishing in the regulations. The wording, unchanged since 1894, allowed Indians to "catch fish for the purpose of providing food for themselves and their families but for no other purpose." The magistrate suggested that an enterprising lawyer might argue that this allowed Indians to catch and sell fish in order to buy food, a plausible reading of the section.[21] (McGregor Young, legal counsel for the recently constituted Royal Commission on Indian Affairs for the Province of British Columbia, which had just begun hearings, came to a similar conclusion, described in Chapter 8.) Fisheries responded that the inspector would simply revoke the food fishing permit if selling fish to procure other foodstuffs became a practice, but in early 1915 it also amended the regulations: "Indians may, at any time, with the permission of the Chief Inspector of Fisheries, catch fish to be used as food for themselves and their families, but for no other purpose."[22] The food fishery was to be only a fishery for food, and Fisheries placed two additional officers on the Fraser that year to enforce it.[23]

In 1917, in what amounted to a significant extension of authority, Fisheries amended the regulations to provide its officials with greater control. The long preamble to the order-in-council explained the "great difficulty" in preventing Indians from selling salmon caught under the auspices of the food fishery. It also lamented the low probability of a conviction, even when fishers were caught and charged.[24] The new provisions, reproduced below, allowed officials to fix the places, methods, and timing of the food fishery. Furthermore, if they could establish that an Indian had sold fish, the burden fell on the accused to establish that the fish had not been caught under a food fishing licence, a reverse-onus provision that enhanced the likelihood of conviction for the sale of food fish. Purchasers of fish caught under a food fishing permit were also guilty of an offence, although it was up to the Crown to establish that they had purchased a food fish.

An Indian may, at any time, with the permission of the chief inspector of fisheries, catch fish to be used as food for himself and his family, but for no other purpose. The chief inspector of fisheries shall have the power in any such permit (a) *to limit or fix the area* of the waters in which such fish may be caught; (b) *to limit or fix means* by which, or the manner in which such fish may be caught; and (c) *to limit or fix the time* in which such permission shall be operative. An Indian shall not fish for or catch fish pursuant to the said permit except in the waters by the means or in the manner and within the time limit expressed in the said permit, and any fish caught pursuant to any such permit shall be deemed to be a violation of these regulations.

(a) *Proof of a sale or of a disposition* by any other means by an Indian of any fish shall be prima facie evidence that such fish was caught by the said Indian, and that it was caught for a purpose other than to be used as food for himself or his family, and *shall throw on the Indian the onus of proving* that such fish was not caught under or pursuant to the provisions of any such permit.

(b) No Indian shall spear, trap or pen fish on their spawning grounds, or in any place leased or set apart for the natural or artificial propagation of fish, or in any other place otherwise specially reserved.

(c) Any person buying any fish or portion of any fish caught under such permit shall be guilty of an offence against these regulations.[25]

These regulations, which gave Fisheries officials new legal tools to restrict the food fishery and further solidified the constructed boundary between food and commercial fishing, were the product of the government's frustration over its inability to contain the food fishery. However, the immediate catalyst may well have been the rock slide at Hells Gate in 1914 and its effect on the Fraser River salmon fishery.[26] In its rush to build track through the Fraser Canyon, the Canadian Northern Railway Company dumped enormous quantities of rock into the river near Hells Gate in 1913. This debris narrowed the canyon and, at certain water levels, increased the water speed and turbulence to the point where the migrating salmon could not get through. Fisheries responded with various strategies, including dynamite to loosen the rocks and wash them downstream. Then in February 1914, a huge portion of the canyon wall at Hells Gate tumbled into the river, changing its contours radically. Fisheries officials scrambled to clear the rubble before the summer sockeye runs, but with mixed success. Some salmon did manage to get through, but they were the fortunate few. In response to the slides, Fisheries officials restricted and then shut the food fishery on the Fraser in 1914.[27] The full extent of the devastation became clearer in 1917, when the sockeye, which live on a four-year cycle, did not return in the expected numbers.[28] Chief Inspector of Fisheries F.H. Cunningham was reported to have said that "the Indian is one of the greatest enemies that the sockeye has."[29] As a result, the food fishery remained closed until 1922 and was heavily restricted thereafter.[30] However, the commercial fishery at the mouth of the Fraser was not restricted, although, as environmental historian Mathew Evenden has pointed out, the reduced runs were a significant factor in the reshaping of the commercial canning industry as it shifted its orientation to the central and northern coasts.[31]

Elsewhere in the province, the evidence suggests that Fisheries officials used the enhanced regulatory powers to increase their enforcement activity. Reverend Peter Kelly, a member of the Haida and president of the Allied Tribes, one of the early Native organizations that had formed to advocate for recognition of Native title in British Columbia, recounted aggressive enforcement practices in a 1923 meeting with senior officials from Indian Affairs:

Almost every Indian Agent I think can report cases where the Indians have been held up, where fisheries officials have come and summonsed Indians to Court for catching salmon for food. Fines have been paid. I am just thinking of one particular instance in Nanaimo only last year, where two parties have gone up the Nanaimo River and speared salmon for food; they had I think two salmon in the canoe, one was on the beach; and they were before the Magistrate, and they were fined. It was not very much, but they were fined nevertheless. I think they paid something like five or six dollars apiece. But for taking food which they thought they had a perfect right to take. Now that sort of a story can come from all parts of the Province. Our friends from the Fraser River have the same story to tell, where wagons have been confiscated, teams of horses have been confiscated, because they were hauling salmon from one reserve to another – from the river to the reserves.[32]

The regulations limited the food fishery to that which could be consumed by the fisher and his or her family, but the goal of officials within Fisheries was to eliminate that fishery as well if it were possible. Chief Inspector of Fisheries Cunningham told the members of the Royal Commission on Indian Affairs in 1915 that he thought "it would be in the public interest to feed the Indian on white men's food, and don't let them eat fish at all."[33] In 1930, in the first issue of *BC Fisherman,* a magazine for the fishing industry, anglers, and boaters, the following story appeared:

INDIAN SALMON DIET EXPENSIVE
How the Noble Redman "Robs" the Salmon Spawning Grounds
By THE EDITOR
When the noble redman was dispossessed of his ancestral heritage extending from the Great Lakes to the Pacific Coast, by the irresistible march of civilization and progress, one of the agreements which the white man made with him was that, within the confines of his reservation, the Indian should be permitted to hunt and fish without hindrance. And that agreement has been carried out to the letter.

But the Indian treaty was drawn up long before anyone in Canada, or at least in British Columbia, gave any thought to the subject of fish conservation.

But the time has come to effect a change in this agreement. Nobody suggests that the Indian should be divested of his inalienable rights. At the same time, if the salmon fishers and salmon canners of British Columbia are to get the benefit of stringent restrictions placed upon them, and the people of Canada are to benefit by the large sums spent in salmon conservation and research work, the Indian will have to change his diet. Salmon has become an extravagance for him.[34]

The Editor went on to suggest that the "Indian Can Find Work" in the industrial commercial fishery and that if Indians still wanted food fish, well, "Indians like pilchards" and the government should offer pilchards as a substitute for sockeye. These views, common among Fisheries officials, led in the 1930s to experimental shipments of commercially caught and processed fish to the interior of the province with the aim of reducing Native dependence on sockeye and other commercially valuable species. The hope was that Native people would eat these commercial products, allowing Fisheries officials to close the food fishery entirely.[35] In effect, that is what happened on the Capilano River in 1925.

R. v. Charlie (1925)

I began this book with a short description of the October 1925 charges against Domanic Charlie (Photo 6.2) for fishing in the Capilano River with a gaff hook. Charlie, a Squamish hereditary chief who had been fishing on the river where it runs through Squamish Indian Reserve No. 5 Capilano, was acquitted at trial, but convicted on appeal and fined one dollar.[36] Although Charlie had been fishing on the reserve and held an Indian food fishing permit, the Department of Fisheries had closed the river earlier in the year to all fishing except sport fishing with a hook and line. It was a measure of just how confined and limited the Indian food fishery had become.

Governor James Douglas appears to have allotted the Capilano reserve, the site of a Squamish village at the mouth of the Capilano River on the north shore of Burrard Inlet, British Columbia, in the early 1860s (Figure 6.1 and Photo 6.3). The precise date is uncertain, but the Royal Engineers surveyed the site as a government reserve in 1863, about the same time that the first sawmills appeared on Burrard Inlet.[37] The Joint Indian Reserve Commission confirmed a reserve of 444 acres in 1877, but it was subsequently reduced by 3.5 acres to allow for a road right-of-way in 1912 and by 18.5 acres in 1913 for a railway through the reserve. In 1916, the Royal Commission on Indian Affairs for the Province of British Columbia cut off another 130 acres on the west bank of the Capilano River at the request of local officials, who argued that the Squamish were not using that part of the reserve and that it could be usefully developed if it were outside the reserve.[38] The next year, the Harbour Commission acquired the foreshore.[39] In short, what little land had been reserved to the Squamish was being squeezed on all sides.[40]

In 1916, the royal commission listed the principal occupations of those living on the Capilano reserve as "fishing, hunting, gardening, towing, hop-picking, and working for wages." The Squamish had twenty-eight reserves, eleven of which the commission identified as "fishing stations," and a population of approximately 400 in 1916.[41] The numbers on the Capilano reserve were disputed (the Squamish claimed seventy-one people lived on the reserve; the Indian agent suggested forty-two), and so was the extent of personal property (the residents claimed to own two gasoline launches, but the commissioners recorded six). The commissioners described the "Condition of Indians" on the reserve as "industrious and fairly well-to-do," which indicated the commissioners' general approval of the lifestyle and their conclusion that the standard of living on the reserve was somewhat above that of most other Native peoples in the province.

Photo 6.2 Domanic Charlie and his daughter Bertha. | *North Vancouver Museum and Archives, Photograph #3755*

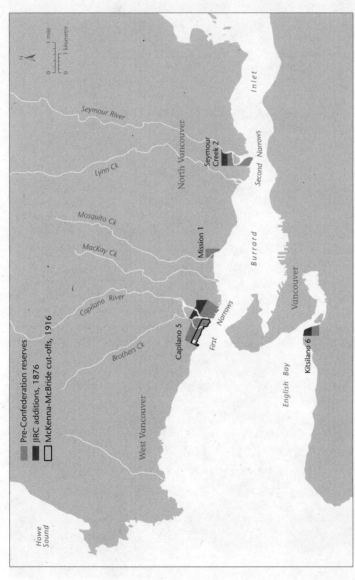

Figure 6.1 Squamish reserves on Burrard Inlet and English Bay in 1925. The Capilano reserve was first allotted in the 1860s. The Joint Indian Reserve Commission added to it in 1876, but the McKenna-McBride Commission cut off 130 acres of valuable waterfront west of the Capilano River in 1916. The first bridge across Burrard Inlet was built at Second Narrows in 1925, the same year Fisheries closed the Indian food fishery on Capilano River and charged Domanic Charlie.

By the early 1920s, the Squamish, one of six bands (including the Sechelt, Sliammon, Homalco, Klahoose, and Musqueam) that Indian Affairs administered through a reconfigured Vancouver Agency, found it increasingly difficult to acquire food fishing permits. In 1922, Fisheries only issued twelve permits to the Squamish. The following year it issued four additional licences, including one to Domanic Charlie.[42] Then, in October 1924, Fisheries officials stopped one of the permit holders, Joe Issac, from fishing with a gaff for salmon in the Capilano River where it ran through the reserve. The chief inspector of Fisheries, J.A. Motherwell, told the Indian agent for the Vancouver Agency, Charles Perry, that because of "the importance of the Capilano River from the standpoint of angling," the river was now closed to any fishing except by rod and hook. He wondered whether it might be "amicably arranged with the Indians to the end that they leave the Capilano alone."[43] Perry conferred with Andrew Paull, who was the secretary of the Squamish Band Council and a central figure in the Allied Tribes.[44] Paull insisted that the Squamish would continue to fish in the Capilano River unless Fisheries could offer some evidence that the closure was needed to conserve the runs and not simply to reallocate fish to anglers. He also wondered under what authority Fisheries had closed the

Photo 6.3 Squamish village at the mouth of the Capilano River, ca. 1912. This view is from Prospect Point, looking north across First Narrows. | *North Vancouver Museum and Archives, Photograph #3653*

river to food fishing while allowing angling.[45] Motherwell responded by closing the Capilano to Natives fishing with nets, gaffs, or anything other than angling gear beginning 1 January 1925.[46] Two more Squamish applied for food fishing permits for the Capilano, but Fisheries refused to provide them.

Fisheries' closure of the Squamish food fishery on the Capilano was part of its general effort on the south coast to enforce the fisheries regulations more vigorously. In August 1925, Fisheries officers apprehended six members of the Homalco and Klahoose peoples for fishing upstream from the fishing boundary in Toba Inlet. The officers confiscated the fish boats and gear, returning them when the season was over, and the accused were fined. At Powell River, Fisheries officers destroyed Sliammon fish weirs.[47] A little farther north, in Bute Inlet, Homalco chief George Harry complained of gendered harassment: "Inside of our reserve No. 2 the Fishery Commissioner don't want any womens in this Reserve to catch any fish for food. Every time a women set a short web to catch a salmon he, the Fishery Commissioner, always pulls the net and takes it away with him."[48] In Bute Inlet, the Provincial Police constable reported that there was "almost a state of war between Indians & the fishery officials" when Fisheries sought to prevent Homalco fishers from fishing in the rivers and selling their catch to cannery boats.[49] The incident led to charges against two Homalco fishers, Tommy Paul and Alex Paul. A justice of the peace (J.P.) in Powell River convicted Tommy Paul of fishing in the river and fined him $102.50 or two months' imprisonment and hard labour in default. Alex Paul received a $50 fine for fishing in a river mouth, $100 for resisting a Fisheries officer, and was ordered to pay an additional $15.75 to cover the costs of the Fisheries officer and the J.P.[50] Both Homalco men ended up in the penitentiary at Oakalla, just outside Vancouver. The police report of the proceedings indicated that "the accused gave evidence in his own behalf and called other Indians as witnesses for the defence, but as they could not understand English, their evidence was not much use." Perry wondered at the "appearance, suddenly, of an almost general attack by regulation enforcement upon the Indians all along the coast."[51]

Through the fall of 1925, Fisheries and Indian Affairs could not agree on who was eligible for food fishing permits. Chief Inspector of Fisheries J.A. Motherwell asked Perry to provide him with a list of names of "deserving Indians" to whom he would grant a food fishing licence. Perry responded promptly with a list of 328 Natives from the six bands under his administration.[52] The list, according to Motherwell, was too expansive: "On a recent inspection trip to the Squamish Reserve an officer of this

Department found only six or seven Indians actually living there, the remainder being scattered – some working at canneries and logging camps and longshoring in Vancouver." Would Perry please send a list of properly deserving Indians, meaning "old and needy Indians" unable to work in the wage economy.[53] Perry stood by his list, which included 126 Squamish in Burrard Inlet and Howe Sound, and castigated Motherwell for the "lamentable ignorance on the part of your office as well as a poor and haphazard evasion of your obvious duty towards the Indians of the coast."[54] Fisheries eventually agreed to reissue food fishing permits to those Squamish who had held them the previous year – by its count, about sixty – but it was determined to enforce the closure of the Capilano River to any fishing other than that with rod and line.

In October, the newly appointed Fisheries officer, Austin Spencer, charged Domanic Charlie for catching two chum salmon with a gaff on the Capilano River. Spencer, an angler himself, frequently fished in the Capilano. In fact, only weeks before the charges, the Capilano had become that much more accessible to sport fishers with the completion of the first bridge across Burrard Inlet, connecting Vancouver to the north shore by road for the first time. Developers extolled the pristine virtues of the newly opened wilderness, and the following description appeared in the *Vancouver Sun* newspaper as part of a twelve-page feature on North Vancouver that coincided with the opening of the bridge: "Within 30 minutes from the time the disciple of Izaak Walton leaves the city of Vancouver he can cast his line at any of at least a hundred spots along the North Shore and seldom does he do so without receiving the answering tug that thrills his heart"[55] (Photo 6.4).

On the Monday following the charges, Charlie appeared with Indian Agent Perry at the Lynn Valley Police Court before Magistrate Alexander Philip.[56] Perry asked for an adjournment, which Philip granted, notwithstanding Spencer's opposition on the grounds that the accused would then have time to prepare a defence. The matter was held over until the following week. In the interim, Perry contacted Fisheries to see if some compromise might be reached. However, Fisheries officials were adamant that the Capilano River would remain closed to all but anglers. Its steelhead made it "one of the best sporting streams in the province," and it was in the "best interests of [the] Province" that the stream should be reserved for angling. Fisheries argued, moreover, that chum were readily available elsewhere, and to close the Capilano would cause the Squamish little hardship.[57]

On 2 November, Charlie returned to court with Perry as well as Andrew Paull and Rev. Peter Kelly, president of the Allied Tribes. Charlie's case was

FISHING ON THE CAPILANO RIVER. NEAR VANCOUVER. B.C·   I. N. Hibben & Co., Victoria, B.C.   No. 11

Photo 6.4   Postcard of sport fishers on the Capilano River, ca. 1905. | *North Vancouver Museum and Archives, Photograph # 8857*

to be an important test case. Spencer led the prosecution, presenting the arguments that Fisheries had exclusive jurisdiction over fishing in rivers and streams, that the department had closed the Capilano River to all fishing except by rod and line, and that Charlie had caught the fish with a gaff. Moreover, Spencer's department had the authority to regulate fishing, including Indian food fishing, under the *Special Fishery Regulations for the Province of British Columbia*.[58] In reply, Perry conceded that Fisheries had jurisdiction to regulate fishing, including the Indian food fisheries, but he argued that the department had exercised its jurisdiction poorly by failing to provide food fishing licences or to inform the Squamish of the restrictions in a timely manner. He also argued that Charlie had been fishing near the mouth of the Capilano River within the Capilano reserve and that Fisheries had no jurisdiction over the activity of Indians on reserves. Furthermore, the *Indian Act* prohibited trespass on Indian reserves, so Fisheries needed the consent of Indian Affairs before issuing angling licences for the waters running through reserves.[59] Magistrate Philip reserved judgment.

In his decision, handed down 12 November, Philip acquitted Charlie. The prosecution had not made out its case against the accused and, more importantly as a general precedent, Fisheries had no jurisdiction over Indians fishing on their reserves.[60] That authority lay exclusively with Indian Affairs. This was an important decision, and a welcome one for Indian

Affairs. The Indian agent at Hazelton wrote to congratulate Perry and to say that he had informed "the followers of Issac Walton," who had complained about Natives fishing from the Moricetown reserve on the Skeena River, that they would be treated as trespassers if he received complaints of their presence on the reserve.[61]

Fisheries was not pleased. It hired A.B. McDonald, K.C., to appeal the decision. Indian Affairs retained J.N. Ellis, K.C., to respond. The Squamish Council thought Charlie should be represented by legal counsel as well and requested that Indian Affairs release funds from their account to hire R.L. Maitland. Indian Affairs refused on the grounds that it had hired Ellis to represent the department as well as the Squamish and Charlie.

The parties appeared before Judge Cayley of the County Court in Vancouver and put to him the question of Fisheries' jurisdiction to regulate Indian fishing on reserves.[62] The lines of argument turned on the constitutional division of powers between Dominion and provincial governments. Fisheries argued that, under the *British North America Act*, it had the sole jurisdiction to regulate the resource, and that it also had a responsibility to enforce its regulations on and off Indian reserves. On the other hand, Indian Affairs appears to have relied on article 13 of the Terms of Union between British Columbia and Canada, which required the Dominion to pursue "a policy as liberal as that hitherto pursued by British Columbia" with respect to Indians.[63] Because the colony had not regulated Native fishing before Confederation, Ellis argued, Native peoples were free to conduct their fisheries without interference.[64] Moreover, the sections of the *Indian Act* that prohibited non-Natives from hunting or fishing on reserve land suggested that Indians had particular privileges on reserves.

Judge Cayley adopted this position in his decision:

> the Indians, as a Band, on each reserve have the exclusive right of fishing on the reserve. No outsider can interfere with them. Penalties are prescribed in the Indian Act if outsiders do interfere. So that the liberal policy pursued prior to Union by the Province of British Columbia is continued by the Dominion Government in accordance with the Act of Union.

Native peoples, therefore, had an exclusive right of fishing on their reserves. But Judge Cayley also held that regulations restricting catching methods did not interfere with that right:

> Are the right, tho, of the Indians as a band interfered with in the slightest degree by prescribing the regulations authorized by the Fisheries Act, as to certain methods of catching fish, allowing some methods and forbidding others? I do not see that

any right whatever of an Indian is interfered with in hunting and fishing on his Reserve. He can hunt and fish all he wishes to. The Fisheries Act only says he shall not use destructive methods.

He went on to state that "there is nothing illiberal whatever in making regulations for the protection of fisheries, whether on a reserve or elsewhere," and therefore "the legal and moral right of the Fisheries Department to pass the regulation in question cannot be disputed." He convicted Charlie, levied a nominal fine of one dollar, and did not award costs.[65]

The Squamish Council asked Indian Affairs to appeal the decision, and Indian Agent Perry recommended appealing as well. In correspondence with Ottawa, he noted that under the *Indian Act,* Indian Affairs had "control and management of the lands and property of the Indians in Canada."[66] Property included fishing rights, and therefore the department should act on its authority to protect Native fisheries.[67] Duncan Campbell Scott, the deputy superintendent general of Indian Affairs, declined the appeal. He felt the management of the fishery would be jeopardized if Fisheries did not have jurisdiction over the entire fishery, including Indian fishing on reserves. Moreover, Scott suggested that Fisheries had never contested the exclusive right of Indians to fish on waters running through their reserves. It was a curious position given that Fisheries had licensed non-Native anglers to fish anywhere on the Capilano. Scott suggested, however, that the Squamish could control the fishing within their reserve, and he recommended that they pass a resolution to regulate and license the non-Native fishers for a reasonable fee.[68] Later that year, the Squamish Council passed a resolution asking Indian Affairs to arrange for anglers licences that would be available, for the nominal sum of one dollar per year, "to white people and persons or Indians other than Indians of the Squamish Band, for fishing in waters passing through any of the reserves of the Squamish Band."[69] Indian Affairs authorized Perry, by letter, to put a licensing system in place, and by the end of October, Perry had issued forty-nine one-dollar licences to residents of Vancouver and North Vancouver who wished to fish in the Capilano River.[70] The following year, Indian Affairs issued at least thirty-five licences.[71] It was not much income to replace the loss of a fishery, and what there was disappeared into the Squamish trust account administered by Indian Affairs.

Having won the Capilano River, Fisheries turned its attention to other streams on the north shore of Burrard Inlet, proposing to close Seymour and Lynn creeks to gaffing as well. Perry remained perplexed by the closures. Anglers were after steelhead, not chum, so why not allow the Squamish to gaff chum? Perhaps the presence of Indians gaffing chum would disrupt the ambient pleasure of angling. As the chum migration was about to begin in September 1926, Fisheries indicated that the Squamish could not use gaffs in the Capilano. Gaffs would only be permitted on four much smaller creeks – Mission, Mosquito, McKay, and Brothers – but not on the Capilano.[72] Perry responded that he would not endorse the closures and could not counsel the Squamish to refrain from gaffing chum. The decision in the Charlie case confirmed the Indians' right to fish, he argued, subject to reasonable regulation, but this did not allow Fisheries to close the river to Indian fishing.[73] Paull wrote that "the closing of the Capilano river is an autocratic ruling and will benifit no one only deprive the Indians of a good part of their nourishment ... and I charge the Government is depriving the Indians of sufficient fish with an attempt to bring about the gradual extinction of the Indians."[74] His goal, and that of the Allied Tribes – to secure recognition of Native title and fishing rights – seemed further away than ever.

Although the Squamish did use the other creeks, they were insufficient without access to the Capilano as well. Each family, according to Paull, needed about 250 dried salmon and 75 to 100 salted salmon between September and April. Chum only migrated up the Capilano and Seymour (which was now closed as well), and they were the fish most easily preserved. The other creeks supported much smaller coho runs. Fisheries Officer Spencer reported that Mission, McKay, and Mosquito creeks, together, would yield between 600 and 700 coho, and Brothers Creek would provide ample chum. The coho estimate, however, was hardly enough to support two families, and, according to Paull, few if any chum migrated up Brothers Creek.[75]

The refusal to allow gaffing in the Capilano was only part of a larger policy to contain Native fishing. When Domanic Charlie applied to Fisheries for a trolling licence, which cost one dollar and was available to anyone, he was refused. It was only when Indian Agent Perry accompanied him to the Fisheries offices that he was sold a licence. Perry noted this as a further example of the general discrimination against Indians within the Department of Fisheries.[76] In Bute Inlet, Fisheries continued to confiscate fishing gear, including valuable fishing nets, from the mouth of the Homathco River.[77] Earlier that year, Fisheries had decided to limit Indian food fishing to a spear and gaff fishery. Nets could only be used on the Nass, Skeena, Nimpkish, and Fraser rivers and on Babine Lake.[78] The Homathco River, however, was too muddy for spears or gaffs; the Department of Fisheries had effectively closed the food fishery.

Conclusion

The Dominion and provincial governments had constructed a pattern of Indian reserve allotments in British Columbia that was premised on Native access to the fisheries. By 1900, however, it was clear that the Dominion had also established a regulatory apparatus intended to minimize Native peoples' access to those fisheries. It did so through the construction of an Indian food fishery that was for the Native fisheries what the reserve allotment process was for Native land. The land policy ignored questions of Native title, confining Native peoples to small reserves intended, at best, to support local subsistence economies; the fisheries policy ignored Native fishing rights and the question of ownership, limiting Native access to an increasingly circumscribed food fishery. Instead of negotiating, the state assumed control of the fisheries and the right to determine who had access. Indian food fisheries, invented and imposed, were conducted at the discretion of the inspector of Fisheries.

Natives' access to their fisheries became a privilege, not a right. This mirrored the province's denial of Native title, to which the Dominion acquiesced. Reserves were to be allotted as needed and rescinded when unused. Many were closely connected to the fisheries, but Fisheries officials repudiated these connections, even where reserve commissioners explicitly assigned exclusive fisheries to the inhabitants of the reserve.[79] At least one critic of my earlier work suggested that in focusing on the loss of Native control over the fisheries, I understated the opportunities for, and the participation of, Native peoples in the industrial commercial fishery.[80] To speak of a Native fishery beyond a food fishery, it was argued, was to miss the fact that Native fishers held and hold a significant proportion of the licences in the commercial fleet. In principle, the commercial fishery was open to all and its rules were to apply to everyone. In fact, there were many injustices. The degree of Native access to the industrial commercial fishery is the focus of the next chapter.

7
Licensing the Commercial Salmon Fishery

From its beginnings on the Fraser River in the early 1870s, the canning industry grew rapidly, if unevenly, to become a major component of the British Columbia economy. It drew hundreds of thousands of investment dollars into the province, employed thousands of workers, and shipped fish worth millions of dollars across the continent and around the British Empire. By the first decades of the twentieth century, fifty to eighty canneries operated each year, many of them isolated industrial work camps on the Nass and Skeena rivers in the north, and around Rivers Inlet on the central coast. The largest cluster occupied the stretch of the Fraser River downstream from New Westminster to Steveston, not far from the emerging city of Vancouver. The fisheries in British Columbia had become the largest and most lucrative in Canada, driven primarily by the industrial commercial complex that caught, canned, and shipped sockeye salmon.[1]

Under the *British North America Act,* "Sea-Coast and Inland Fisheries" were a Dominion responsibility.[2] As a result, when British Columbia joined Confederation in 1871, management of the fisheries passed into the hands of the Department of Marine and Fisheries (Fisheries), based in Ottawa, just as the first canneries were appearing on the Fraser. Initially slow to establish the department's presence – Canada's *Fisheries Act* did not come into force in British Columbia until 1877 – Fisheries officials had already heard calls from cannery owners to regulate the industry. The canners' concern was less the preservation of fish in the face of scarcity, although the upstream activities of Indians and gold miners were thought to be damaging the runs, than it was about limiting competition by restricting new entrants to the industry or managing labour through control of the fishing licences. However, with few officers in the field and a tenuous understanding of the fisheries in British Columbia, the Dominion governed the commercial fishery with a light hand in the early decades of the industry. With a few important exceptions, the Dominion constructed the fisheries as an open-access resource, placing some restrictions on fishing gear, opening times, and fishing locations, but generally allowing anyone to purchase a licence and, if they could outfit themselves with a boat and nets, join the fishery. The foundation of this open-access regime lay in the common-law doctrine of the public right to fish, and its principle that the public had a right not to be excluded from the fisheries.[3] The exceptions to open access included an attempt over three seasons (1889-91) to institute a limited-licence regime on the Fraser, putting most licences under the direct control of the canneries, and a similar, although more sustained, attempt to create a limited-licence regime (1911-19) on the central and northern coasts, again allocating most of the licences to the canneries. Another important departure from this principle of open access lay in the granting of exclusive seine-net licences along much of the coast and a few salmon-trap licences on the south coast. But these exceptions aside – exceptions that had a considerable impact on the development of the fisheries and that are discussed below – the Department of Fisheries constructed an open-access regime in the fisheries, a regime that remained in place until Minister of Fisheries Jack Davis reintroduced limited licences in the salmon fishery in the late 1960s.[4]

If the Department of Fisheries operated with a light hand on the canning industry, effectively opening the resource to those who invested in the processing of fish, it kept a firmer grip on the Native fisheries, confining them to the margins as the industrial commercial fishery grew. As a result, Native peoples' participation in the fisheries depended on the cannery operators' need for their labour, not on their ownership of the resource. The rise of the canning industry did create new opportunities for Native peoples, many of whom participated in the industrial commercial fishery as fishers, cannery workers, and very occasionally as cannery operators.[5] Indeed, Native workers, both men and women, were more involved in the industrial commercial fisheries than in any other industry in the province, and the seasonal work in the canneries or on fish boats became an integral part of many coastal communities. Moreover, Native people's labour played a significant and essential role in the industry, particularly in its early years on the south coast and through the first half of the twentieth century on the north coast.[6] This is to be expected, given that the fisheries had long been at the centre of most Native economies on the Pacific coast. But the refusal to recognize Native ownership of or priority to any portion of the resource meant that Native peoples' participation in the fishery depended on the usefulness of their labour.

Controlling the size and composition of a labour force, Native and non-Native, and therefore the price of labour, was a principal preoccupation of colonial administrators in many corners of the British Empire.[7] In some settler colonies, particularly in South Australia and New Zealand, where the political economist Edward Gibbon Wakefield had considerable influence, colonial officials attempted to use the price of land and the level of immigration as a means to ensure not only an adequate supply of labour for capital (meaning low-cost labour), but also to create a social structure that reflected the aspirations of the dominant class. This meant attempting to attract a white, preferably British, working class, whose members would provide suitable labour and, in less than a generation of industrious activity, amass enough capital to acquire land and become the employers of a new wave of immigrants.[8] These ideas had some purchase in British Columbia, particularly on Vancouver Island, but the proximity of the United States, where land could be had more cheaply, limited their effectiveness. Nonetheless, the goal of a white British Columbia mirrored the aims elsewhere in the empire, most notoriously, perhaps, the White Australia policy.

Fuelled by the racism of the late nineteenth and early twentieth centuries, the provincial and Dominion governments designed policies to encourage white immigrants but discourage Asians. To this end, the state imposed head taxes on Chinese immigrants, limited their participation in the democratic process, and restricted their access to important sectors of the economy.[9] As an emerging power in the Pacific, Japan was able to exercise its international influence to spare its nationals some of the overtly racist policies, and Japanese fishers came to comprise a majority of the fishing fleets, but they were targets nonetheless. As the number of Japanese fishers grew, so did the demands and the policies to limit their presence in the industry. In areas where fish were abundant and agricultural opportunities limited, fishing licences became one of the principal tools to structure not only the desired racial composition of the fishing fleet but also the patterns of settlement more generally. Native fishers were affected by these policies, sometimes to their advantage, as when Japanese fishers were forced from the industry. However, the gains were always temporary, offset by policies to encourage white settlement and by the smothering imposition of the *Indian Act,* which constructed Native peoples as wards of the state.

The result was a state-structured racialization of the labour force, with policies designed to privilege white fishers and the owners of capital.[10] Native fishers who sought to participate in the industrial commercial fishery faced many obstacles, not the least of which was a lack of access to capital to finance boats and fishing gear. The degree to which these obstructions limited the ability of Native fishers to work as commercial fishers requires further investigation, more than I am able to devote here. On this front,

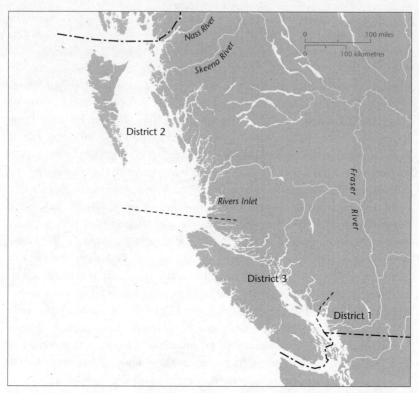

Figure 7.1 Department of Marine and Fisheries districts. In 1904, the department divided the management of the fisheries into northern and southern districts. The following year it separated the Fraser River fishery from the southern district, creating a third district encompassing the waters surrounding Vancouver Island.

my comments are suggestive rather than conclusive. In this chapter I am primarily concerned with the policies of the Dominion government, sometimes transferred into law, to regulate the fishing fleet that supplied the salmon canning industry, and the impact of these rules on Native participation in the industrial commercial fisheries. The chapter is organized by gear type and region, reflecting the different management regimes for each and the differing impact on Native fishers (Figure 7.1).

The provinces, by dint of their jurisdiction over property, had begun in the late nineteenth century to assert a right to manage the fisheries, precipitating a series of cases to demarcate control of the resource.[11] For its part, British Columbia passed a *Fisheries Act* in 1901 and created its own Department of Fisheries to carve out some jurisdictional space and to collect revenue from the fisheries.[12] The eventual outcome of this dispute was a lopsided division of jurisdiction in favour of the Dominion,[13] but various provincial regulatory regimes affected the commercial fisheries, particularly where the fishery was conducted from or required access to the foreshore. Those developments are touched on in this chapter, but the focus remains on the work of the Dominion Department of Fisheries.

Drift-Net Licences on the Fraser River (District 1)

The commercial salmon fishery in nineteenth- and early-twentieth-century British Columbia was predominantly a drift-net fishery targeting migrating sockeye (Photo 7.1). At the mouths of British Columbia's major rivers, fishers deployed gill nets from hundreds of skiffs that had been towed by a cannery-owned tender to the fishing grounds.[14] Using nets suspended from a cork line on the surface and weighted with a lead line to hang down in the water, and working in pairs from a skiff, a fisher who handled the nets and a boat puller who worked the oars would stretch a net across a river (under the regulations, the net could be no more than one-third the river's width) or near its mouth. With one end of the net attached to a buoy, the other to the skiff, the assemblage would drift with the current, the boat puller keeping the net perpendicular to the path of the migrating salmon. As the salmon headed into the river or upstream, they would swim into the mesh. Unable to squeeze their bodies through the openings or to back out because of their gills, they would be caught. At the end of a "drift," the fishers would haul the net aboard, disentangle the fish, and row back upstream for another set or to the cannery or tender to unload their catch. In 1890, the Dominion superintendent of Fish Culture, Samuel Wilmot, described the drift-net fishery that year:

The "Saw-kay" [sockeye] salmon enter the Fraser in July. They are principally caught with drift nets. The regulation length of these nets is 150 fathoms, and size of mesh 5¾ inches extension measure. These nets are cast from the fishermen's boats and allowed to drift with the tide on the surface, being kept in place by the usual cork and lead system. The salmon passing up come in contact with these nets and get "gilled." The fishing boats, which are limited in number on this river to 500, are seen dotted all along the river. By far the greater amount of fishing is carried on near the outlet of the river into the straits, where the fishermen seem most anxious to set their nets, for the first chance in meeting the incoming fish from the sea. It is not unusual for a single boat, in one drift of a net of a quarter mile, to fill the boat with several hundred salmon. These are immediately taken to the company's factory where the boat is owned or the fisherman is employed, and thrown upon the landing platform.[15]

Photo 7.1   Drift-net fishing on the Fraser River, 1906. The fisher stands in the prow, minding the net, while the boat puller works the oars. | *Provincial Archives of Alberta, Charles Mathers Collection, B9852*

The commercial fleet grew quickly in the early years to supply the burgeoning canning industry. In an effort to establish its authority and secure a small revenue stream, beginning in 1881 the Department of Fisheries required that all fishers on the Fraser who were catching salmon for the canneries near its mouth purchase an annual licence.[16] Nearly all of the 600 to 700 annual licence holders on the Fraser from 1882 to 1887 were Native, an indication of the degree to which the canneries relied on Native labour in the early years of the industry.[17]

The rapid expansion of cannery capacity and of the number of licensees began to raise questions about the sustainability of the fishery and the profitability of the industry. A massive fishing effort and severe habitat destruction had already produced sharp declines in the salmon runs in California and on the Columbia River, and some Fisheries officers in British Columbia began to raise concerns about the Fraser.[18] Cannery operators also expressed concerns, although primarily about the rapid expansion of the industry and their ability to secure enough salmon given the increasing competition. At times a fractious group, in competition over fish, workers, and markets, the cannery operators at other times formed a tight-knit cabal to control labour or to present a common front to the government (Photo 7.2). These men sought restrictions and greater control over the licences, primarily to limit new entrants in the industry. In 1889, Fisheries complied.

Photo 7.2   Salmon canners on the Fraser River, 1889. Left to right: (seated) D.J. Munn, E.A. Wadhams, Alexander Ewan, M.M. English, Ben Young; (standing) Mike Leary, H. Harlock, T.E. Ladner, J.A. Laidlaw, Robert Matherson. | *Delta Museum and Archives/MSS-DE1982-14-1(P)*

It instituted a limited-licence regime, restricting the number of salmon drift-net licences on the Fraser to 450. The canneries would divide 350 licences between them, and the remaining 100 would go to independent fishers.[19] The limited-licence regime induced a flood of applicants for the independent licences, but few new applicants received licences, given that there were many more than 100 previously licensed independent fishers.[20] Not surprisingly, the new regime provoked uproar among the independent fishers, and in 1890, Fisheries increased the number of independent licences to 150 and the total number of licences to 500.[21] The same limits remained in effect the following year; then all were removed in 1892.

The limited-licence regime affected many fishers, but it had a dramatic effect on Native commercial fishers, cutting their number in half, from more than 600 in each of the six years preceding the restrictions to a few more than 300 during the three-year experiment in limiting licences.[22] Of these, most worked under a cannery licence. According to the testimony of Captain George of Chehalis to the British Columbia Fishery Commission, 1892, only forty Natives held independent licences in 1891, fewer than one-third of the 150 independent licences, although many more had boats and nets and wished to fish independently.[23] And while some Native communities had fewer independent licences than they wished, others were excluded altogether. When Cowichan fishers applied for independent licences on the Fraser in 1889 and 1890, Fisheries denied them on the grounds that they were not members of a resident tribe.[24] The Cowichan's winter villages and Indian reserves straddled the Cowichan River on Vancouver Island, but every summer they crossed the Strait of Georgia to fish sockeye near the mouth of the Fraser River. This seasonal round long predated the canning industry and had continued until the 1890s as Cowichan fishers took advantage of the new market for their fish. However, the site of their fishing village on the Fraser had been sold to a white settler before the Indian reserve commission was established, and the reserve commissioners did not allot an alternate site. Without a reserve on the Fraser, Fisheries did not consider them a resident tribe and would not allocate independent fishing licences to Cowichan fishers.[25]

Fisheries confirmed the end of the short-lived limited-licence regime on the Fraser in 1894, when it introduced regulations stipulating that "each *bonâ fide* fisherman, being an actual resident of the province of British Columbia," could obtain one salmon fishing licence.[26] Otherwise, the number of commercial licences that fish processors could hold varied roughly with the amount of capital invested: canners were entitled to twenty licences; exporters and wholesalers of fresh, salted, smoked, or cured salmon could have seven. Each commercial licence cost ten dollars. In comparison, a "domestic licence," permitting the holder to fish for food, cost one dollar.

With the end of the limited-licence regime, the number of licensees grew dramatically, and the makeup of the fishing fleet began to change in two important respects. First, newly issued independent licences accounted for most of the increase in licensees, so that by the late 1890s, only a small minority of fishers on the Fraser worked under cannery licences.[27] Fisheries confirmed this transition in 1899 when it reduced the number of licences a cannery could hold to ten. The canners, it appears, were not particularly concerned by this reduction in cannery licences; many of the fishers who held independent licences were beholden to a cannery for pre-season credit to outfit their boats, or even for the boats themselves. Many fishers, nominally independent, only fished for the cannery to which they were indebted. Henry Bell-Irving, owner of the Anglo-British Columbia Packing Co., the largest canning company in British Columbia through the 1890s, claimed that by 1900, each cannery deployed as many as 250 boats, vastly exceeding what they had been allowed under the licence restrictions ten years earlier.[28]

The ethnic composition of the Fraser fishing fleet was also fundamentally transformed. From the lifting of the licence restrictions for the 1892 season until 1905, roughly 500 Native fishers held annual licences to fish commercially on the Fraser. Over this same period, the number of Japanese and white fishers grew dramatically. In most years there were well over 1,000 licensed Japanese fishers (in some years as many as 1,800) and nearly 1,000 white fishers. Based on these figures, Native fishers made up, on average, 44 percent of the Fraser fishery from 1891 to 1894, a figure that fell to 12 percent between 1901 and 1904.[29]

In response to the growing Japanese presence in the fleet, Fisheries, in 1899, introduced a salmon fishers' registry and the requirement that independent fishers be British subjects. All independent licence applicants, except Native applicants, had to register their name and address with Fisheries before they were eligible to receive a licence.[30] In addition, fishers of Japanese ancestry and "foreigners" who did not speak English had to produce naturalization papers to prove they were British subjects in order to join the registry.[31] In a further effort to control the composition of the fleet, Fisheries modified the regulations in 1900 to establish a boat pullers' registry and to require that all boat pullers register. Indian boat pullers, who were frequently the wives or female partners of fishers (Photo 7.3), were exempt. The intent and effect of these registries were to create obstacles for Japanese fishers, although certainly not insurmountable ones; they, or the canneries on their behalf, produced naturalization papers and received licences in growing numbers.

The number of Native fishers, which had declined as a proportion of the fleet in the 1890s, declined absolutely in the first two decades of the twentieth century. In those years, Japanese fishers made up the majority of the fleet on the Fraser, followed by white and then Native fishers. These racial divisions were important markers in the industry and were played upon by the cannery owners to dampen the force of the frequent labour unrest in the years around 1900.[32] As categories to describe the nature of the fishing fleet, however, they are problematic, concealing great diversity within the named groups, particularly within the "white" category, whose members shared little more in common than that they were immigrants who were not from Japan. Fisheries recorded the nationality of fishers on the Fraser in 1915 (set out in Table 7.1). It reveals a diverse fishing fleet, but also a declining Native presence in commercial boats on the Fraser.

By 1920, Native participation in the Fraser River commercial fleet was almost insignificant. Over the 1920s and early 1930s, the number of Native licence holders varied from twenty-six to seventy-three, in a fleet that fluctuated between 1,000 and 1,500 licence holders.[33] On several occasions,

Photo 7.3 Native fisher and boat puller, likely husband and wife, working their Columbia River-style gill-net boat on the Fraser River, 1913. In the background, a cannery tender moves a barge used for collecting salmon from the fish boats. | *BC Archives, F. Dundas Todd Collection, HP84113*

Table 7.1

Nationality of Fraser River fishing licence holders, 1915

Nationality	Number	Nationality	Number
Canadian	286	Austrian	42
English	139	American	11
Scotch	90	Hollander	1
Indian	295	Chilian	4
Greek	53	Italian	8
Norwegian	143	Russian	12
Swede	42	Dane	4
Jap	1,320	French	5
German	9	Turk	1
Finlander	47	Hawaian	1
French Canadian	17	Philippino	1
Icelander	3	Pole	1
Irish	15	Welsh	1
Spanish	25	Maltese	1

Whites 962, Indians 295, Japs 1,320

Note: The nationalities are identified as per the source.
Source: F.H. Cunningham, chief inspector of Fisheries, to D.N. McIntyre, provincial deputy commissioner of Fisheries, 26 August 1915, GR 435, box 18, file 157, BC Archives.

the Department of Indian Affairs sought to arrest the decline with requests that Fisheries issue commercial licences to Native fishers at no charge, to no avail.[34] The Japanese presence began to decline as well in the 1920s, a result of specific policies to reduce Asian participation in the fishery.[35] By the early 1930s, the commercial fleet on the Fraser was primarily white (perhaps better described as non-Native and non-Japanese), with a significant although reduced Japanese presence, and very few Native fishers.

In Washington State's Puget Sound, site of another substantial fishery targeting Fraser River sockeye, the decline of Native participation in the commercial salmon fishery occurred earlier and more precipitously. According to anthropologist Daniel Boxberger, the Coast Salish Lummis "were almost totally excluded from commercial fishing by 1900, as quickly as they had been incorporated only a few years before."[36] He explains that this was one of the outcomes of a technological change in the fishery. In the 1890s, the canneries began using salmon traps, also known as pound nets, to catch the majority of the sockeye heading through US waters to the Fraser River. These were massive, capital-intensive operations that extended almost a kilometre from shore, creating a fence and a series of impoundments to trap the migrating salmon (Figure 7.3). Although expensive to construct, once built they required little labour and were so efficient that they rapidly displaced other fishing methods. The salmon traps also physically displaced important Indian reef-net fisheries by occupying either the reef-net fishing ground or the area immediately adjacent to it.[37] The Indian labour that had briefly been important became irrelevant, and the Lummi did not have access to the capital necessary to build a salmon trap.

There is not such a clear explanation for the decline in the number of Native licence holders on the Fraser. Part of the answer, I think, lies beyond the fishing industry and its regulation. At one level, the apparatus of the Canadian state, including the *Indian Act* and residential schools, was encircling and disrupting Native lives, detaching individuals from family, languages, and traditions, and interposing itself between Native peoples and everyone else. The reserves were becoming places in which to contain the Native presence in British Columbia, erasing it from the rest of the province, rather than bases from which to engage with the larger society and economy. On another level, cannery operators drew on a growing number of immigrants in the towns of New Westminster and Steveston and the emerging city of Vancouver, many of whom were more than happy to have seasonal work in the salmon fishery. Even though the licences were nominally independent, cannery operators frequently owned the boats and the nets or provided significant pre-season credit for this equipment to those fishers whom they wished to hire. In 1913, when the Royal Commission on Indian Affairs for the Province of British Columbia visited the Squamish at the Capilano reserve in North Vancouver, Chief Mathias Joseph testified that a fish boat cost $150, nets between $125 and $150, and a gill-net licence $10. When the sockeye runs peaked, a good fisher might clear as much as $800, but in other years he might not be able to cover debts.[38] Gasoline-powered boats, which entered the south coast fishery in increasing numbers in the 1910s, would have doubled or tripled the cost of a boat, adding significantly to the initial outlay required to enter the fishery. The ten-dollar licence fee, in place since 1894, was relatively small, but symbolically significant. Many Natives resented paying for permission to fish in waters they had long considered theirs. After 1908, fishers had to buy five-dollar licences from both the Dominion and the province, and this only aggravated their sense of injustice. Musqueam chief Jonnie complained of their constricting presence to the royal commission: "When I want to go fishing, the two parties are also holding on to each end of my boat – There are initials and numbers on the bow and initials and numbers on the stern, and I know that I own the water that is the grievance that I want to bring before the Commissioners. I don't want to have a licence to do anything. When I want to catch fish for my living I don't want to be interfered with at all."[39]

Japanese immigrants, many from fishing communities in Japan, had quickly established themselves as industrious and effective fishers and came to predominate, notwithstanding the pervasive racism directed against them. Only targeted government policy would significantly reduce their participation in the fishing fleet on the Fraser. Employing Japanese fishers was uncluttered and uncomplicated in comparison to hiring Native fishers, whose employment and ability to acquire debt were made difficult by their status as wards of the state under the *Indian Act*,[40] and who had the suppressed but unresolved questions of Native title and rights to the fisheries lingering in the background. Although a much fuller study is needed, these were the factors that appear to have contributed to the steady decline of Native fishers' participation in the commercial fishery on the Fraser over the first two decades of the twentieth century.

Drift-Net Licences on the Central and Northern Coasts (District 2)
The state regulation of the salmon fishing industry began on the Fraser in the 1870s and spread from there to the Skeena and Nass rivers on the north coast, Rivers Inlet on the central coast, and to other canning centres as they appeared. Initially, a uniform set of regulations applied everywhere,

but this began to change with the limited-licence regime that Fisheries imposed on the Fraser River in the late 1880s. After that short-lived experiment, the Fraser came to be fished by fishers holding at least nominally independent licences. On the central and northern coasts – Fisheries District 2 – cannery control of licences remained a feature of the fishery into the 1920s.

In the first decade of the twentieth century, the canners in the north sought to consolidate their control by shutting out independent fishers. They did not wait for the Department of Fisheries to intervene. In 1910, the four canners on the Nass agreed not to purchase fish from independent fishers. Instead, they would limit themselves to sixty-five boats each, divide the 120 Indian crews equally, pay ten cents for sockeye (prices were stipulated for other fish as well), rent their fish boats for two dollars per week, pay twenty cents per hour for work inside the cannery cleaning fish and four cents for packing a twenty-four-tin flat of one-pound tins, limit net length to 200 fathoms, and pay no bonuses "to any chiefs[,] individuals or fishermen or for Indian labour."[41] The agreement was a private attempt to manage the fisheries in the absence of effective state regulation, and it underscores the level of collusion facing fishers and cannery workers, Native and non-Native.[42]

The composition of the fishing fleet on the Skeena River in 1909 reveals the cannery control of the northern fleet. The ten canneries operating at or near the mouth of the river that year employed 798 fishers, only five of whom held independent licences.[43] There was also a much stronger Native presence in the north coast fleet in the first quarter of the twentieth century than there was on the Fraser. Of the 798 fishers, 399 were Japanese, 349 Indian, 49 white, and 1 was "coloured." White fishers held the few independent licences.[44]

Having devised informal arrangements to manage the fisheries, the cannery operators on the central and northern coasts sought to formalize their control of the fishing licences in the Dominion's fishing regulations. They pushed for a boat-rating scheme that would limit the number of boats on each of the major rivers and divide them between the canneries. Fisheries delivered such a scheme for the fishing season in 1911: there would be 850 boats on the Skeena, 700 at Rivers Inlet, 240 on the Nass, and smaller numbers on other, more minor, rivers and inlets.[45] In total, Fisheries issued 1,990 gill-net licences, 16 drag-seine licences, 2 purse-seine licences, and 13 unidentified seine licences for District 2 in 1911.

Generally satisfied with a boat-rating scheme that gave them control over fishing licences and limited new entrants, the canners on the north and central coasts expressed great concern when, in 1912, Dominion and provincial fisheries officials agreed to grant independent licences to white fishers. The two levels of government had decided that it was "eminently desirable to have the fisheries in the hands of a suitable class of *white* fishermen" and that, in order to spur white immigration, independent white fishers should have the first priority for licences.[46] By the early years of the twentieth century, few white settlers had arrived and stayed on British Columbia's central and northern coasts. Province-wide, slightly more than half of the 50,000 inhabitants of British Columbia counted in the 1881 census were identified as Indians.[47] Of the 3,086 people listed as living in the lower Skeena and Nass region, virtually all of them were identified as Indians (2,893 or 94 percent). Only 92 whites and 101 Chinese lived in the area. Various proposals had been floated over the years to increase the white presence, including a plan in the 1880s to encourage the immigration of displaced Scottish crofters, who were thought to be hardy, industrious workers and capable fishers.[48] This and other schemes had not materialized. The number of white inhabitants on the central and north coasts increased slowly, but the idea of populating the region with white settlers remained.

Negotiations between Fisheries and the cannery operators produced an agreement that, beginning in 1913, Fisheries would reserve 20 percent of the licences on the north coast as independent licences for "*bona fide* white fishermen owning their own boats and gear."[49] However, to ensure that Native fishers were not shut out, Fisheries instructed the cannery operators to hire Native fishers first under the cannery licences. In an effort to compel compliance, the department indicated that it would not issue licences to a cannery unless it had arranged to employ "local Indians."[50]

Some cannery operators were leery of any change that removed licences from their control, even if they supported efforts to increase white settlement.[51] Speaking on behalf of most of the northern canners and in defence of the boat-rating scheme as initially construed, Henry Bell-Irving, chair of the Anglo-British Columbia Packing Co., painted a picture for Fisheries officials of the progress and industry that the cannery operator brought to the wilderness:

> Imagine a fiord in Northern British Columbia running deep into the interior, between high mountains, the slopes of which are clothed with dense forest. No white man's habitation within a hundred miles, no human beings in the neighbourhood, except a few families of Indians, but at the proper season millions of Salmon. The cannery-man arrives, and breaks the solitude. He clears a space ashore, erects his wharf and factory, and dwellings for his men, instals [*sic*] modern

machinery and brings up materials, boats and nets. He then scours the country for labour, and soon what was once a solitude, is a busy hive of industry. His salmon pack is put up, representing a value of possibly $100,000 or more, and fish which before his arrival were chiefly food for the bears and eagles, are converted into a valuable article of commerce. Later on the politician appears on the scene ... He says to the canneryman – "It is true you have staked a lot of money in erecting and equipping your factory, but you must understand that after this the fish belong to the fishermen only, and you must buy fish from them at their price, though you will have the privilege as before of equipping them with boats and nets, and of advancing them cash and supplies when necessary." The cannerymen thinks this is scarcely a square deal, but patiently continues. The large sums annually distributed in wages are of substantial benefit to the country in all branches of trade.[52]

It was a Lockean argument. The lands lay in waste, without improvement and with little or no benefit to humans, until the cannery man organized capital and labour to put them to productive use. This use, which benefited all, created the basis of the cannery man's claim to ownership, or at least to secure and stable access to the fishery. Native peoples, on the other hand, were assumed to be part of the wilderness, without any better claim to the fish than the bears or eagles.[53]

Many canners were concerned that the drive to populate the coast with whites would limit their ability to attract labour. They generally supported the goal of keeping "Orientals" off the north coast, but in order to alleviate concerns about the supply of labour, the Prince Rupert Conservative Association passed a resolution asking that "Indians be considered as whites in granting of licenses."[54] This view was not unanimously held. British Columbia Packers Association (BC Packers), which by 1913 controlled a large portion of the fishing industry in the province, offered the following advice to the minister:

The issue of fishing licenses to Indians is especially to be deprecated. They are wards of the Dominion Government, and specially treated and protected. The Canners have always treated them with the utmost consideration, and provided them with houses, food, medicines etc., and their wives and daughters have worked in the Canneries, and made 15¢ to 25¢ an hour.

It is not deemed advisable to grant Indians "Independent" licenses, as they are liable to mis-interpret the reason and become difficult to manage by the authorities.[55]

In this view, Native peoples were wards of the state and, as such, had received special treatment from the state and from the canneries themselves.

Of course, the canners rented the houses and sold the provisions as part of their businesses; fishers and cannery workers had earned their accommodation and food. The apprehension that Indians might become "difficult to manage" if they held independent licences was perhaps more accurately a concern that the price of fish might rise if Fisheries issued independent licences. The feared "misinterpretation" suggests a concern that Native peoples might understand the independent licences as a step towards the confirmation of their fishing rights.

Much of the concern over labour shortages focused on Native women's labour. Skilled and relatively inexpensive, Native women cannery workers had become at least as valuable to the canning industry in the north as Native fishers.[56] In time, the Japanese fishers provided an alternative fishing fleet. Many canners used Chinese labour inside the canneries, but immigration restrictions reduced the supply. As a result, there was not a reliable, skilled, and inexpensive alternative to Native women's work in the canneries. The problem for the canners, as their member of parliament put it, was that "it is impossible to get the Indian women without employing their husbands," and many of the "husbands" did not want to work under cannery licences.[57] One company in Prince Rupert, Canadian Fish and Cold Storage, applied for and secured thirty independent licences for Native fishers on the grounds that it could not operate without the labour of Native women, and it could only attract that labour if there were independent licences for Native men. This application provoked scorn and derision from some cannery operators, who claimed to manage full well with white labour.[58] Most were sympathetic to the difficulties of procuring labour but angrily denounced the company for circumventing the boat-rating scheme.[59] Fisheries granted the licences, but the provincial Department of Fisheries, which was seeking to assert its authority over aspects of the commercial fisheries, revoked the company's cannery licence. It had been issued, wrote the provincial commissioner of Fisheries, on the grounds that "only white fishermen and only white labour would be employed in connection with the cannery."[60]

At first, white fishers were slow to apply for the independent licences. In 1913, Fisheries issued 175 independent licences to whites for the central and north coasts, only slightly more than the 170 licences that had been set aside for white fishers on the Skeena alone. In 1914, however, Fisheries issued 456 independent licences, and 575 the following year.[61] Anthropologist John Pritchard described the race for licences as "something of a farce."[62] White fishers registered land pre-emptions in order to establish their residency and acquire a licence, but they had no intention of staying longer than the fishing season. The authors of the 1917 Fishery Commission Report dubbed them "raft-farmers" because they did little more than build and live on a temporary raft adjacent to their pre-emption.[63] Cannery operators, intent on securing their supply of salmon, scrambled to attract these licence holders whether or not they had boats or fishing gear. This race to attract the independent white fishers, Pritchard suggests, squeezed out Native fishers.

As employees without independent licences, Native fishers were in a precarious position. The wages from fishing and cannery work had become an important component of many coastal Native communities. Work in the salmon fishery was seasonal and cyclical, with years of plenty followed by relative dearth. While these cycles had always been a feature of the salmon fishery, pre-industrial economies had been more diversified than the industrial commercial fishery that relied almost exclusively on salmon.[64] When the salmon runs were large, there was plenty of employment, but many of the canneries and the jobs associated with them disappeared in lean years. The canneries, furthermore, exercised considerable control over the fishers who worked under cannery licences, often in cannery boats and with cannery nets. The disciplining effect of the power to withhold the means to participate in the fishery is difficult to quantify, and the power of the canneries certainly varied with their need for labour. But for people who, through the imposition of Indian reserves, the *Indian Act,* and Canadian fisheries law, had come to rely on cannery work, the prospect that a cannery operator might withhold licences loomed over their participation in the workforce. A letter from Tinshop George, a chief of the Oweekeno from Rivers Inlet, to the provincial premier, Richard McBride, reveals the anguish caused when, after a dispute over the number of licences, a new cannery manager withheld licences from the Oweekeno:

My people is in greatest trouble in our life. As not one of them get gear to catch fish this summer ...

The trouble is going by the new manager, the successor to G.S. McTavish. Mr. McTavish use to gave all the people of the Oweekaynoes their gears who wishes to fish here in previous season but we are in greatest trouble. We believe that it is not right to not suppo[rt] the Aborigines of Oweekayno River and we say since the first cannery planted in our richest place. And you can fairly well understand that most of my people are in middle of age and more older they only get their wages in fishing only. But after all that they could not get either job until another season come again. And it seems to us that we might get no more money to buy food or some of the people got crowd of family which I believe is going to be greatest trouble so I pray gave my humbly petition to you to try and help us out of trouble as we are your poorest Indian obedient friends.[65]

In 1913, Tsimshian and Gitxsan fishers, many of whom owned boats and nets and had fished commercially for years, wrote to the minister of Fisheries, J.D. Hazen, to protest the department's refusal to grant them independent licences:

> We the undersigned Indians of Northern British Columbia, hereby Respectfully ask you that we be allowed to use or purchase Independent Salmon fishing Licences. During the present season we were told by the Fishery Officials of your Department that we could not purchase an Independent Licence as it was only for white men who owned their boat and Net.
>
> We have fished on the various rivers of Northern British Columbia, for a good many years, and many of us have our own boats and fishing gear, and we think we are entitled to have an independent licence to catch salmon, provided we comply with regulations laid down by your Department.
>
> We are natives of this Country, and as Fishing is one of our means of livelihood, and we are loyal British subjects, we think it only right and fair that if we have the money to purchase a Licence, and the other qualifications necessary, we be allowed to have these independent Licences.[66]

If Fisheries responded, there is no record of it. The Port Simpson Tsimshian sent a similar petition to the provincial government later that year, pointing out that they had fish boats and gear, that independent fishers received considerably better prices for their fish than those working under a cannery licence, and that they wished the opportunity to purchase independent licences.[67] There is no record of a response to this petition either, but a memo from the deputy commissioner of Fisheries reveals the department's policy: "It was felt that to grant these licences indiscriminately to Indians would defeat the object of the policy, – to introduce the settlement of white men in the north, since the Indians as wards of the Government have exceptional advantages, and the total allotment of licences for this purpose would be taken out by Indians."[68] The deputy commissioner and the chief inspector of Fisheries also believed "that the Indian would never be satisfactory as a free fisherman as he was less reliable and would be apt to break contract and leave the employing canners in the lurch."[69]

In 1914, a group of Tsimshian fishers hired a law firm to plead their case. In an argument similar to that which Fisheries had earlier used to deny exclusive Native fisheries, the lawyers pointed out that there was no provision in the regulations or statutes to allow Fisheries officers to discriminate when it issued fishing licences. What were the legal grounds on which it was creating an exclusive white fishery? Fisheries avoided the question, responding that there was no shortage of employment opportunities for

Native fishers and that white settlement on the north coast was a priority.[70] When Tsimshian spokesperson Benjamin Bennet addressed the Royal Commission on Indian Affairs for the Province of British Columbia in 1915, this was his response:

> When the Indians found out that the cannery men would not give the Indians a chance to fish we sent a petition down to Ottawa and we told Ottawa "We are in a bad fix – the cannery men won't let us fish for them, and what are we going to do about it – we want independent licenses, we want to be able to catch fish and sell the fish for the highest price just like the white people," and the answer came back "You are living on a Reserve – you cannot get an independent license – you are not a voter. Only white men can get independent licences – you are living on a reserve and you cannot get an independent license – you are under the Indian Act, and that Act will attend to your wants." This reserve is no good to us, and because we are living on a reserve we cannot make any money – we are under the Indian Act.[71]

The commissioners listened. In their separate, confidential report, which accompanied the official report in 1916 on the status of Indian reserves, the commissioners included a strong condemnation of the "class or racial discrimination" in the government policy to issue independent licences only to whites on the north coast.

> The Commission is unanimously of opinion that the Indians of Northern British Columbia are – but should not be – discriminated against in the issuance and use of these "independent" fishing licenses; and that there is no authority conferred by the law, or intent therein expressed or suggested, for such class or racial discrimination. The Commission is of the opinion that in the matter of "independent" fishing licenses, applications of Northern British Columbia Indians should (as are applications therefor of white fishermen and of Indian fishermen on the Fraser River) be considered and dealt with upon their individual merits and not refused because of the applicant being an Indian, the Indians of British Columbia being British subjects and as such entitled to equal consideration with their fellow British subjects.[72]

The commissioners also reported that Native fishers had difficulty acquiring cannery licences. Canneries preferred hiring Japanese fishers and did so notwithstanding a government directive that these licences should go first to Native fishers. Without access to independent licences, and with limited access to cannery licences, which Fisheries hoped to phase out as more white settlers took up independent licences, the commissioners reported that Native fishers would "in a few years be completely s[h]ut off in the North from the salmon fishing industry."[73] Native fishers, in their testimony to the subsequent Fishery Commission of 1917, concurred:

> The Indians always get smaller licences every year. We want to find out how many licences we are supposed to be getting. They are always getting less every year, licences for us. We are liable to not fish in some years. There is very few Indians fishing now to-day compared with before.
>
> Q. But do they want to fish?
>
> A. Sure they want to fish, but lots of fellows just stay home; can't get no licence; because I think we have a right over any people, because we have no other chance.[74]

The report of the 1917 Fishery Commission concluded that it was a laudable goal to encourage the participation of "native born Canadians, other than Indians," in the fishing industry, but tying licences to settlement was ineffective. The commissioners also stated, in a mild critique of the government's white-only policy as it affected other immigrants, that "it is equally desirable that every class that comes to Canada to make permanent homes here and become citizens, should have a proportionate opportunity of engaging in every kind of honest, useful work. The blending of the various elements of the population in work is the best way to bring about real assimilation." Their critique of the government's exclusion of Native fishers was stronger. Although they would not countenance what they described as the "extreme claim" of Native peoples to a prior right, they did suggest that Native fishers should have priority to more than a food fishery:

> If disproportionate consideration is to be given to any class, we feel that it should be to the Indians. The position and point of view of the Indians was placed before us with ability and eloquence by Indian spokesmen. While we cannot agree with the extreme claim to a sort of prior right to do commercial fishing, in view of the fact that the Indians can do now all they could before the advent of the white man, that is, take whatever fish they can use themselves, nevertheless the tastes of the Indians have been becoming more diversified, and they need many things which can only be secured with money and the opportunities on the northern coast for earning money are very limited except in commercial fishing. All the Indians who by steadiness and skill can qualify for a certificate of competency may well be given an opportunity out of proportion to other elements of the population, but the standard should be applied in their case just as in that of others.[75]

The commissioners hedged their recommendation with qualifications, but they appear to have advocated something akin to a right to a moderate livelihood from the fishery, not just a food fishery.

These opinions notwithstanding, Fisheries continued to deny Native peoples access to independent licences on the central and north coasts until 1919.[76] Beginning that year, Native fishers could acquire independent drift-net licences. Seine-net and salmon-trap licences, the focus of the following section, as well as licences to operate a cannery, were still only available to whites and would remain so until 1924. And even if Natives were now eligible to hold independent gill-net licences, the goal of the Dominion and provincial governments remained that "a white fishing population should be permanently established on the Pacific Coast to engage in the great fisheries thereof."[77]

Seine Nets, Salmon Traps, and the Foreshore
Drift-net fishing dominated the salmon industry in late-nineteenth- and early-twentieth-century British Columbia, particularly on the Fraser and other major salmon rivers. Elsewhere in the province, fishers turned to other technologies, principally seine nets and salmon traps. While generally allowing drift nets anywhere in tidal waters, except near the mouths of all but the largest rivers, Fisheries maintained closer scrutiny of the seines and traps, at times prohibiting them entirely in the salmon fishery and, in the case of salmon traps, limiting their use to a few locations.

Large boats with power winches began to appear in the late 1920s, but the early seine-net fisheries were relatively small operations that relied primarily on hand-hauled nets. They caught a small proportion of the salmon taken in British Columbia but were a highly effective fishing technology on many of the province's smaller rivers. Salmon would congregate in sheltered bays near river mouths along the coast, waiting for the rising water levels that came with the fall rains and allowed them to ascend the river to their spawning grounds. In a drag-seine fishery, one end of the seine net was attached to land and the other was towed out in a semi-circle and then back to land, entrapping the fish. The net, suspended from a cork line and weighted with a lead line, was then hauled to shore, sometimes with a donkey engine, bringing the fish with it (Figure 7.2). The purse seine was deployed on the water. Using two boats, fishers would play the net out in a circle around a school of fish and then draw it shut at the bottom. They would haul the net aboard or tow it to land.[78] Skilfully set, seine nets could catch most of a small salmon run before it entered the river.

In his 1890 report on the British Columbia fisheries, Samuel Wilmot suggested that it was difficult to gauge the extent of seine fishing in British Columbia and, furthermore, that the regulations were unclear as to whether it was permitted. Fisheries had, however, issued a few licences, and their effects were quickly felt. The first complaints were heard in 1888, when

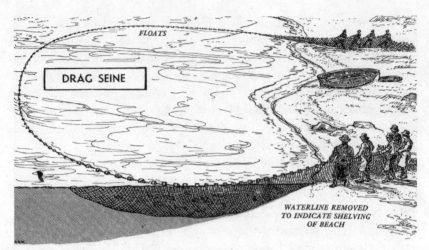

Figure 7.2   Diagram of drag-seine fishery. In some locations, the fishers used donkey engines to pull the nets ashore. | *Source*: Frederick William Wallace, "The Story of Canada's Fishing Industry," *Canadian Fisheries Manual* (1942), 128.

the Cowichan Tribes on the east coast of Vancouver Island protested that boats with seine nets were drawing fish out of Cowichan Bay, taking them before they entered the Cowichan River, and leaving the Cowichan weirs empty.[79] Fisheries did not rescind the licences, but it did move the fishing boundary a short distance out into Cowichan Bay, making it somewhat more difficult for net fishers to catch the schooling fish.[80]

The seine licences in Cowichan Bay were among the few that Fisheries had issued, and following Wilmot's recommendation to limit the salmon fishery in British Columbia to a drift-net fishery, Fisheries prohibited the use of seine nets to catch salmon following the 1890 season.[81] In 1891, Fisheries prohibited the use of purse seines anywhere in Canada.[82] The prohibition notwithstanding, Fisheries continued to issue between six and nine salmon seine licences each year through the 1890s.[83] In 1890, it issued six seine-net licences, including two to the Lowe Inlet Canning Company, which operated on the north coast in Kitkatla territory, just south of the Skeena. It reissued these licences until 1894, when Captain J. Walbran of the *Quadra*, a steamer that patrolled the west coast, charged John Rood, manager of the Lowe Inlet cannery, for deploying seine nets in the inlet and surrounding waters. New regulations in 1894 were contradictory. One section prohibited the use of nets in the salmon fishery except drift nets deployed in tidal waters. Another section stipulated that seine nets could not be used within 500 yards of the river mouths, suggesting that they might be used beyond this distance.[84] Rood admitted using the seines, fished by Native fishers, but seemed perplexed by the charges. Fisheries had licensed the cannery to fish with at least two seine nets in Lowe Inlet since 1890, and apparently Rood thought he could continue after the amendments in 1894. In Provincial Court he was convicted and fined $480.[85] The fines seem to have had little impact on the company's production, and in 1897, Fisheries issued six seine licences to the company for fishing in Lowe Inlet.[86]

Seine nets reappeared elsewhere in the province as well, including Cowichan Bay, producing more protests from the Cowichan.[87] Given this pattern of licensing, Fisheries officials in British Columbia appear to have decided, prohibition notwithstanding, that seine nets could be deployed along the coast, but not to fish the sockeye runs on rivers such as the Fraser, Skeena, and Nass. Localized seine-net fisheries would be allowed, particularly for the less valuable coho and chum salmon. The licences, however, were not available to everyone. The seine-net licences went only to white fishers and primarily to cannery operators. In 1899, Fisheries refused a Cowichan request for a purse-seine licence to catch chum in Cowichan Bay while continuing to allow a non-Native fleet of gill-netters and drag-seiners to operate there.[88] Then in 1902, in a letter denying a request from one of the tribes on the Saanich Peninsula to use a net on Pender Island, the deputy minister of Fisheries provided the following general policy statement:

> Seine nets have been granted to white fishermen only, one of the conditions being that the seine licensee should be running a cannery or under obligation to erect one. Hence Indians have received hitherto gill-net or drift-net licences only. It would be very unfair to grant such important fishing privileges to Indians and refuse white fishermen, who were in the same position of not being able to erect and run canneries. Indians of course are allowed much freedom when fishing for food and for their own use; but when fishing commercially in competition with white fishermen exceptional privileges cannot be allowed them. Indians as a rule cannot be relied on to abide by strict rules, and were seine and reef net privileges granted to them serious abuses would immediately arise.[89]

Indians could not be trusted to fish within the rules, or so maintained a Department of Fisheries that, earlier in the decade, had continued to issue seine licences to white cannery operators in contravention of its own prohibition.

In 1903, the Department of Indian Affairs asked Fisheries to stop issuing seine-net licences to non-Native fishers operating near the mouths of the rivers on which reserves had been allotted in Sliammon, Homalco, and Klahoose territory north of Vancouver. The reserves had been allotted, the deputy superintendent general of Indian Affairs argued, "for the very purpose of insuring them an ample supply of salmon," and the seine-net fishing threatened that supply. Fisheries replied that it did not issue licences without first considering the interests of the Indians, and in this case it did not see a conflict between the Indian food fishery, which focused on chum, and the commercial seine operations that targeted sockeye.[90] On occasion, Fisheries did protect Native fisheries by refusing non-Native requests for fishing licences when the Native fishery coincided with other non-Native interests. The Cowichan were able to forestall several cannery requests for licences in Cowichan Bay, in part because of the river's reputation among sport fishers. Fisheries also denied a request for an exclusive lease on the Salmon River (Vancouver Island) because it supplied a logging camp upriver and the Komox, who had a reserve at its mouth.[91] Generally, however, Fisheries ignored Native interests when allocating licences; its officials certainly thought themselves under no obligation to preserve Native access.

Unlike the small-scale, labour-intensive seine fisheries, salmon traps were massive, capital-intensive operations. Nets were fixed to pilings that extended hundreds of metres out from the shoreline, perpendicular to the land (Figure 7.3). These nets operated as long fences, channelling the migrating salmon into a small impoundment from which they could not escape. Used widely in US waters around the San Juan Islands and at Point Roberts, where they caught a large proportion of the sockeye heading for the Fraser, salmon traps were generally prohibited in British Columbia except in the few places where they might intercept fish before they entered US waters. This included Boundary Bay, where Fisheries issued two licences in the 1890s to allow fishers to intercept the sockeye before they rounded Point Roberts,[92] and just west of Victoria at Sooke, where traps built in the early 1900s prevented some fish from ending up in US nets deployed from the San Juan Islands.[93] The nets also limited the catch on adjacent Indian reserves. When the first salmon traps were proposed at Sooke, Indian Superintendent and Reserve Commissioner A.W. Vowell reminded a Fisheries official of the connection between Indian reserves and the fisheries: "The Indians do not object to fish traps so long as these do not interfere with their fishing arrangements but it may be remembered that when their Reserves were located by the Indian Reserve Commission they were promised to the effect that their ancient fishing rights would in no-wise be interfered with."[94]

Although they operated at very different scales, in most cases salmon-trap and seine-net fisheries required access to the foreshore. This meant

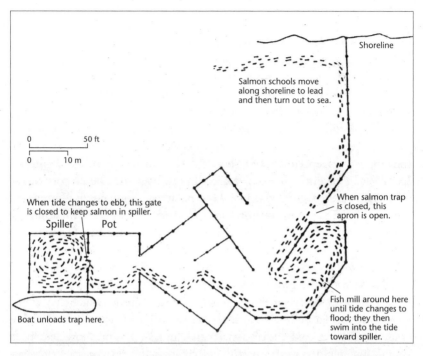

Figures in diagram:

Shoreline

Salmon schools move along shoreline to lead and then turn out to sea.

0   50 ft
0   10 m

When tide changes to ebb, this gate is closed to keep salmon in spiller.

Spiller   Pot

When salmon trap is closed, this apron is open.

Boat unloads trap here.

Fish mill around here until tide changes to flood; they then swim into the tide toward spiller.

Figure 7.3   Diagram of the Diamond Salmon Trap at Sooke. | *Source*: Adapted from sketch by Frank Gray in the "The Pot and Spiller" (1977), Sooke Region Museum.

acquiring a provincial lease. Since 1894, the province's *Land Act* had permitted the chief commissioner of Lands and Works to issue twenty-one-year leases for fishing stations.[95] In 1900, and perhaps earlier, the provincial government began to receive applications for foreshore leases, primarily around the southern tip of Vancouver Island, where land-based fishing technology could be deployed to catch the migrating sockeye as they headed for the Fraser River. By 1903, the province approved a standard form for foreshore leases, providing a term of five years at a rate of $100 a year.[96] The following year it issued a call for tenders, indicating that it would grant foreshore leases to whoever offered the largest cash bonus to the province in addition to the yearly rent.[97] With the two levels of government involved in issuing licences for different aspects of the seine-net and salmon-trap fisheries, conflict erupted when different people held different licences to occupy the same spaces.[98] The question of jurisdiction simmered for a few years, but the eventual compromise seems to have been that the province and the Dominion would coordinate their leases and licences so that a fisher using seine nets or a salmon trap would have both a Dominion fishing licence and a provincial foreshore lease.[99] The licences were geographically specific and exclusive, allowing only the licence holder to fish in the inlet or bay, or along the stretch of coastline.

What was the status of foreshore fronting Indian reserves? If it were part of the reserve, the Dominion held it in trust for the Indians, and the province had no jurisdiction. If it were not within the reserve, it was land that the province could lease as any other. This was the province's position, and the Dominion Department of Fisheries concurred. Indian Affairs disagreed. In 1879, in an effort to resolve the dispute over the sale of part of the Cowichan reserve on Cowichan Bay, Indian Superintendent Powell had written of the value of the foreshore to reserve communities along the coast: "It may be that I am unable to comprehend what 'the foreshore' of an Indian reserve is. I am however quite sure of the fact that the water frontage of all coast reserves constitutes the most valuable and useful part of them so far as the Indians are concerned, and to be deprived of such an important portion of their allotment would be an injustice which they would never willingly submit to."[100] The Department of Justice sided with Fisheries. In 1905 it told Indian Affairs that unless the foreshore had been specifically included in the instrument creating the reserve, the foreshore was not a part of the reserve despite its value to the community: "The Indians have no title, nor has the Crown any title in trust for them, to the foreshore adjoining an Indian reserve. The foreshore may be a part of a reserve, and whether it is so or not depends on the instrument creating or constituting the reserve, but if the reserve does not include the lands forming the foreshore, the ownership of the land adjoining carries with it no title to those lands."[101] To confirm this opinion and assert its jurisdiction, the province amended its land legislation to provide that no foreshore was part of an Indian reserve unless the province also approved the land grant.[102]

In 1904, Fisheries sent an officer to investigate a disturbance in Jervis Inlet. A drag-seine operator was hauling nets across the foreshore of a Sechelt reserve and had built a saltery on the reserve. The officer reported that in the past the Sechelt had allowed the fishery as long as the holder of the licence, Mr. Murray, employed them to do the work. This was one of the many local compromises that probably suited both parties for a time, but in 1904, Murray refused to hire Sechelt fishers, and they refused to let him use the foreshore unless he paid them. There was no other land in the vicinity suitable for hauling drag seines from the water, so the Fisheries officer proposed that Murray take a purse-seine licence so he could catch the fish and then tow the nets to another landing area off the reserve. Fisheries approved this resolution, which ended Murray's encroachment on the reserve but allowed the fishery to continue in the waters immediately adjacent to it.[103]

The question of ownership of reserve foreshores remained unresolved and a serious concern in many coastal communities. Among the demands of the Allied Indian Tribes of British Columbia in response to the report of the Royal Commission on Indian Affairs was the insistence "that all foreshores whether tidal or inland be included in the reserves with which they are connected, so that the various Tribes shall have full permanent and beneficial title to such foreshores."[104]

Expanding the Seine-Net Fisheries, 1904-24

In 1904, Fisheries amended the *Fisheries Regulations for the Province of British Columbia* to allow, but also to limit and regulate, the seine-net and salmon-trap fisheries.[105] It then began issuing seine licences for many areas along the coast, entitling the holders to exclusive seine-net fisheries for identified stretches of the coast. Although the seine fisheries, which targeted the many lesser runs of salmon, remained a small fraction of the drift-net fishery, they frequently occupied waters adjacent to a reserve, including those set aside by the reserve commission for fishing. The runs on these rivers and creeks might number from a few thousand fish to several tens of thousands, and a seine or two working in the bay could catch most of these fish, dramatically reducing the numbers available to the Native peoples who lived on the reserve. The use of seines in these circumstances prompted repeated complaints.

In 1905, the Heiltsuk protested the grant of seine licences to non-Heiltsuk fisher Robert Draney. The licences affected the fisheries associated with several Heiltsuk reserves that the Heiltsuk understood had been reserved to them when the Indian reserves were allotted. Draney also hauled some of his drag seines ashore on reserve land. If Draney were to be allowed to continue fishing, the Heiltsuk argued, he should pay the Heiltsuk the drag-seine licence fee of twenty-five dollars per year for each stream.[106] It was not just the Heiltsuk who were concerned. Indian Superintendent and Reserve Commissioner Vowell explained that,

> as regards the claim made by the Bella Bella [Heiltsuk] and other Indians to the license fee paid by Mr. Draney, the Indians contend that as Mr. Draney fishes in tidal water at the mouths of the small streams on which their fishing stations are located, he catches the fish that would otherwise ascend into their traps, and which they consequently consider belong to them. This would be the case also with the Tsimpsean, Kitkathla, Kitimat, Kitkata, and Kitlope streams anyone netting within half a mile of the mouth of the said stream would monopolize nearly the whole of

the fish that otherwise might reasonably be supposed to go up those streams to their spawning beds.[107]

Fisheries refused to reassign the licences or to remit the licence fee to the Heiltsuk. Draney, Fisheries officials noted, hired Heiltsuk fishers; the department would therefore not prohibit him from landing his nets on reserve land.[108] In effect, the Heiltsuk could sell their labour but not their fish. However, when the Royal Commission on Indian Affairs visited Bella Bella in 1913, Draney Fisheries Ltd. was hiring Japanese fishers to fish its drag- and purse-seine licences.[109] By 1915, the company held six drag-seine and three purse-seine licences in Heiltsuk territory. Heiltsuk fishers held no seine licences, and their labour in the industrial commercial fishery seemed increasingly threatened as well (Figure 7.4).

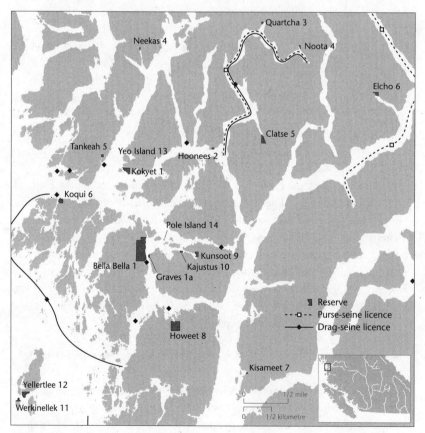

Figure 7.4   Heiltsuk reserves and approximate locations of some Draney Fisheries Ltd. seine-net licences, 1915. | *Source*: Licence locations as described in Cunningham to McIntyre, 20 August 1915, GR 435, box 41, file 367, BCA.

Where seine licences had not already been issued, Fisheries seemed willing to consider Native applicants, at least for a time. In 1906, two Heiltsuk fishers, Peter Starr and Jacob White, applied for drag-seine licences to operate near the mouths of several rivers on the central coast. Inspector J.T. Williams noted that drag-seine licences had not yet been issued for the areas requested, but neither had such licences been issued to Native fishers before. Did he have authority to do so? Fisheries officials in Ottawa assured him that he did.[110] The Metlakatla received a drag-seine licence in 1906 or shortly thereafter.[111]

However, Fisheries continued to issue seine licences in waters adjacent to coastal reserves and the window of opportunity for Native applicants was short. In 1907, the provincial government amended the *Land Act* to prohibit the sale of Crown land to "aborigines of this continent,"[112] and the Dominion government appears to have followed a similar approach to issuing seine-net and salmon-trap licences: only whites were eligible. In 1911, the Kitkatla again expressed concern about their lack of access to the licences for Lowe Inlet. The creeks in and surrounding Lowe Inlet, and the fisheries in those creeks, were owned, they said, and not for Fisheries to allocate to others:

> We, that is the people of this village, are sending you this letter. You probably know that we fish creeks around here for salmon, and wish to buy licenses for these creeks. We have always fished them, long before any cannery was built here.
>
> The following own creeks of their own:
>
> Moses Gladstone
> Robert Brown
> John Davis
> Oswald Tolsnie
> George Macaulay
> Alfred Robinson
> Frederick Gladstone
> Amos Collison
>
> Each of these wants his own license and will pay for it. Do not give the licenses to Lowe Inlet Cannery or Mr. Curtis; our money is just as good as theirs, therefore we ask you for these licenses.
>
> The license business has just become clear to us. We have always been blinded by the Canneries in connection with licenses. We have worked these creeks for Lowe Inlet ever since there was a cannery there and we worked them before there was a cannery.

Upon receipt of this letter kindly send us an early reply. We cannot work these creeks any more if we have no license. We catch all the fish and have always done so. The creeks belong to us and we ask you to hand us the licenses.[113]

Ignoring the statement of Native ownership, Inspector of Fisheries F.H. Cunningham replied that nobody could fish commercially above high tide, that the Lowe Inlet Canning Company had held the fishing licences in the adjacent waters for over twenty years, that the Kitkatla (including those who had applied) fished under these licences, and that they were "in a much better position at the present time than they had been before the advent of the white man."[114] They had plenty of employment, and the canneries did not interfere with their food fishery. They had, therefore, nothing to complain about.

The following year, the Kincolith, through the Indian agent C.C. Perry, expressed concerns about seine nets adjacent to reserves north of the Nass River that had been allotted to them specifically as fishing stations:

I beg to report that the Indians of the Kincolith tribe have, through their council, raised a strong objection to the use of drag seines for fishing of salmon by Japanese and others in the several creeks which have been fished from for food by the Indians for many years; such creeks being situate in Observatory Inlet, including Hastings Arm and Alice Arm, and being in the vicinity of the Indians' fishing stations.[115]

The Japanese fishers were operating under cannery licences, as cannery employees. Fisheries' position was that it would not issue drag-seine licences to the Kincolith, but it would do what it could to convince the canneries to employ Native rather than Japanese fishers.[116] The Kincolith were not satisfied. Fisheries had, by then, issued two seine licences to the Metlakatla, the only such licences held by Native peoples, so why would it not issue licences to them? The obstacle, Perry explained, was the economic interest of the canneries:

The main objection of cannery men to the further issuance of licenses to Indians is, it seems, a fear lest the Indians holding such licenses should ask a higher price for their fish than the canners would be disposed to pay. Conditions are such that, should an Indian ask 25 cents per fish whilst holding a license and other Indians operating for the canners and without individual licenses are paid 15 cents, there might be serious discontent amongst the Indians, and there would likely be a clamouring on the part of the Indians for a monopoly of licenses, which monopoly would now appear to be in the hands of the canners.[117]

In the 1915-16 season, Fisheries issued twenty purse-seine, eighty-one drag-seine, and two trap licences for District 2 (central and north coasts). In District 3 (west coast of Vancouver Island and Johnstone Strait) it had issued forty-one purse-seine, thirty-three drag-seine, and ten trap licences.[118] Canning companies held most of the licences, which entitled the holder to an exclusive seine-net or trap fishery in the waters described in the licence. In some cases, Fisheries also refused to issue gill-net licences in order to protect the exclusive rights of the seine licensee. In response to a 1915 query from the Indian agent about gill-net licences for Natives in Knight Inlet, Fisheries responded that it would not issue those licences in waters set aside for seines.[119] In fact, Fisheries had issued seine licences covering a relatively small area of Knight Inlet, but much of the rest of the coast was blanketed with exclusive seine fisheries.

The extent of the exclusive seine-net and salmon-trap fisheries is revealed in Figures 7.5 to 7.7. The maps show the reserves allotted specifically for fishing purposes along several stretches of British Columbia's coastline. They also indicate the seine-net and trap licences that Fisheries issued for the 1915-16 fishing season in those same coastal waters. Each seine licensee held the exclusive right to fish commercially in the areas identified in the licence, and the mapping reveals just how much of the coast had been allocated as exclusive fishery for the canneries. The Metlakatla held the only licence issued to Native fishers, but many of the licences permitted holders to fish in waters adjacent to reserves that had been allotted to secure access to local fisheries. A great many of the reserves that secured access to a salmon fishery now had a seine-net licence holder on the doorstep, sometimes even using and occupying reserve foreshore. To point to one of innumerable examples, BC Packers held two seine licences for the Nimpkish River, giving the firm prior access to the 'Namgis peoples' fisheries and greatly reducing the runs to the three reserves (Ches-la-kee 3, Ar-ce-wy-ee 4, and O-tsaw-las 5) that Reserve Commissioner Peter O'Reilly had allotted in 1886 to secure access to those runs (see inset map, Figure 7.6).

The following exchange in 1914 between 'Namgis chief Legeuse and Commissioner McKenna of the royal commission reveals the impact of these licences:

> Commissioner McKenna: You said you used to sell fish, but you could not sell them now. Do you mean that you were stopped, or do you mean that there is no market outside of the cannery that is here?
>
> Chief Legeuse: I mean that years ago when I was a young man, they were able to use the traps all along to get the salmon and we sold them from these traps; but

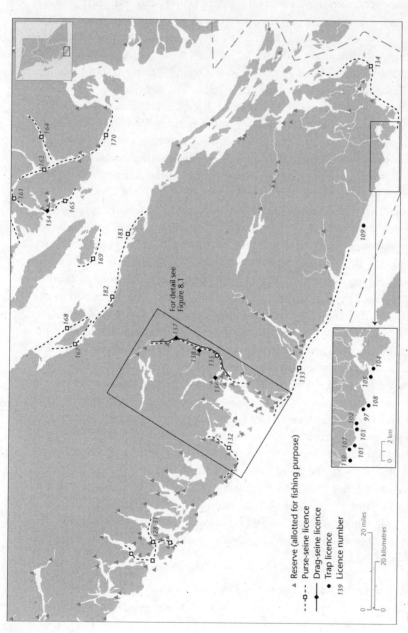

Legend:
- ▲ Reserve (allotted for fishing purpose)
- ▫--- Purse-seine licence
- ◆— Drag-seine licence
- ● Trap licence
- 139 Licence number

For detail see
Figure 8.1

Figure 7.5 Indian reserves allotted for fishing purposes and the areas reserved for commercial seine-net and salmon-trap licences for the 1915-16 fishing season in the southern portion of District 3. | *Source.* Licence locations in this and the following maps as described in Cunningham to McIntyre, 20 August 1915, GR 435, box 41, file 367, BCA.

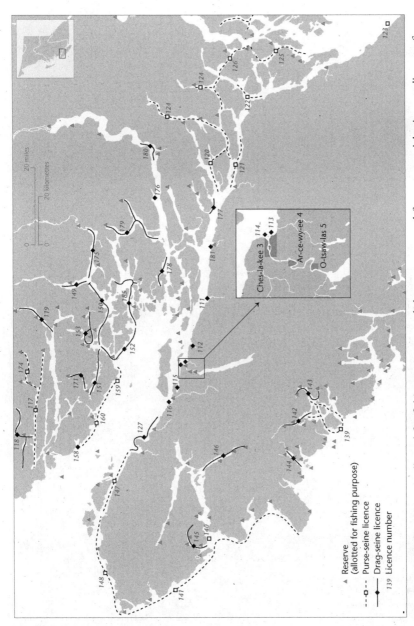

Figure 7.6 Indian reserves allotted for fishing purposes and the areas reserved for commercial seine-net licences for the 1915-16 fishing season in the northern portion of District 3.

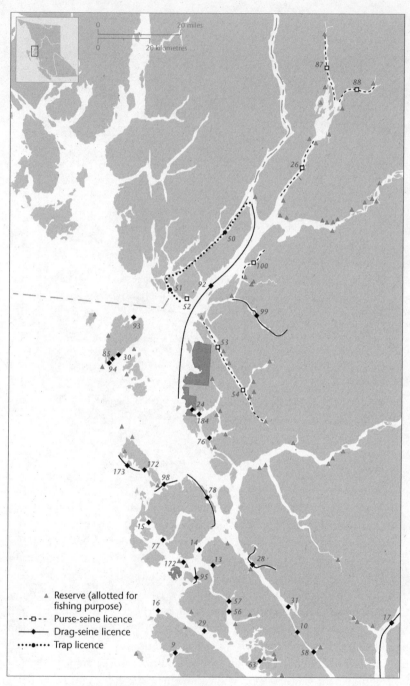

Figure 7.7 Indian reserves allotted for fishing purposes and the areas reserved for commercial seine-net and salmon-trap licences for the 1915-16 fishing season in the northern portion of District 2.

now we are not allowed to use the traps and sell them from the traps. The sockeye salmon we have been stopped from selling the fish at any time of the year.

Q. Can you catch them for your own food?

A. We are not allowed now because we are not allowed the traps.

Q. And it is because they have not the traps now that they don't catch them – Do you refer particularly to the Nimpkish river?

A. Yes, that is my river. Our old trap grounds are in this river.

Q. And you are not allowed to catch any salmon in that river for sale?

A. No, because the cannery has been given the exclusive right to catch fish in that river.

Q. Are you not allowed to catch sockeyes in that river for food?

A. We cannot use our traps, and we cannot get it from the men that use it for commercial purposes. The canneries are fishing with the nets and they won't let us have any of the fish in the nets.

Q. That river empties into the Straits through your Reserve No. 5 – Are the Indians not allowed to catch sockeyes in that part of the stream that runs through that Reserve?

A. No, we are not allowed.

Q. When the salmon have been running, have they been stopped catching the salmon with nets in that part of the river that runs through their Reserve?

A. We don't and are not allowed. We wish we could.

Q. Were you ever fishing there with a net and a whiteman came along and stopped you?

A. Yes, I have been stopped.

Q. Do you know the name of the man?

A. The cannery stopped us.[120]

In some cases, Native fishers worked the cannery-held seine licences, although their employment was far from secure. At a 1923 meeting between senior officials in Indian Affairs and a number of Native leaders, including the executive of the Allied Tribes, Kwakwaka'wakw spokesperson Mrs. Cook described how O'Reilly had allotted the reserves at the mouth of the Nimpkish River because of their value as fishing stations. Cannery fishers, she continued, now occupied the foreshore and the best fishing sites. Native fishers frequently worked those cannery seines, but because they could not hold a seine licence themselves, they had, in effect, been forced to concede their right to the foreshore in exchange for uncertain employment (Photo 7.4). At the same meeting, Reverend Peter Kelly, a member of the Haida and chair of the Allied Indian Tribes, described how Haida fishers in Masset Inlet had "use[d] force" to keep Japanese

Photo 7.4   'Namgis (Alert Bay) drag-seine crew on the Nimpkish River. |
BC Archives D-04636

fishers from working the Wallace Fisheries Ltd. seine licences in the inlet.
However, Kelly reported that in Skidegate Inlet, most of the boats working the cannery licences had Japanese fishers; only two boats had Native
crews in 1923.[121]

Not until it amended the regulations following the 1923 fishing season
would Fisheries again consider issuing seine licences to Native fishers.[122]
This did not mean, of course, that Native fishers received seine licences
adjacent to their reserves or anywhere else. It meant that, beginning in
1924, if Fisheries had not already issued a licence for an area that it considered eligible for a seine licence, or if such a licence became available, Native
fishers were eligible to apply.

Conclusion

Except for an increasingly restricted Indian food fishery, the Dominion
Department of Fisheries refused to recognize any prior right of Native
peoples to the fisheries. It also denied any connection between the allotment of Indian reserves and the fisheries. Instead, Fisheries opened the
resource to the industrial commercial fisheries and, in a few places, to the
emerging sport fisheries. Native people participated in the British Columbia fisheries not as owners of a resource, but only to the extent that their

food fishery did not interfere with canning or sporting interests and to the extent that their labour was useful to those same interests.

Fisheries built its management regime on the principle that fish were common property. In a context of presumed abundance, such as the Fraser River sockeye runs, and in a fishery where the commercial fleet had first access to the migrating salmon, this model suited the cannery operators, who were largely able to dictate the terms on which the industry operated, even if they did not directly hold the fishing licences. However, Fisheries modified the principle of common property when it was in the interests of the canneries to do so – for example, by issuing exclusive area licences for seine fisheries on many of the smaller salmon rivers along the coast, and by establishing a partial limited-licence regime for the drift-net fisheries in the larger rivers on the central and northern coasts. Canning interests held the great majority of these licences.

Native people worked in the industrial commercial fishery as fishers and cannery workers, and in the industry's early days their labour was essential to its success. But as the need for Native labour diminished, so did their access to the fishery. The canneries increasingly turned to a labour force detached from other obligations as well as other means of support, and thus highly dependent on employment. This produced a more malleable, perhaps exploitable, labour force. Some have argued that Native people, in contrast to the Japanese fishers to whom they were frequently compared, retained strong ties to a local territory and kinship connections, and that these ties, which created obligations that had to be attended to, also provided a means of support that lay apart from cannery employment. According to these scholars, this lack of complete dependence on the wage economy and the unreliability associated with it made Native fishers less desirable employees, even if they were no less able as workers.[123] Unlike Japanese fishers, whose continued presence in British Columbia depended entirely on cannery employment and housing,[124] Native workers had not yet been so effectively detached from the pre-industrial economy. And yet, this continued connection to place and to a local subsistence economy meant canners did not have to pay Native workers year-round to maintain a labour force for the seasonal work. If labour could be had more cheaply as a result, one might have expected cannery operators to seek out Native labour, which they did with Native women shoreworkers.[125]

Regional and gender-based variations notwithstanding, Natives participated in this new economy as fishers and cannery workers, but they did so in an environment over which they had little control, under a legal regime that ignored their prior claims to ownership and jurisdiction, and as targets of discrimination and prejudice in the allocation of licences. Even where their labour might have been valued highly, the Dominion and province intervened with policies intended to encourage white settlement, in doing so excluding Native fishers from certain types of licences or from operating canneries. As a result, Native peoples were stuck between the state's refusal to recognize Native rights and title, and its discriminatory allocation of licences, caught in a policy designed to contain their access to the resource. Recent judicial statements that Natives suffered no discrimination in their attempts to participate in the fishery are based on a profound misunderstanding of the fishery and its regulation in the late nineteenth and early twentieth centuries.[126]

8
Land and Fisheries Detached

In this chapter I return to the allotment of Indian reserves in British Columbia and to the work of the royal commission charged with resolving what had become known as "the Indian land question." What the commissioners encountered when they began their hearings in Native communities in 1913 were the consequences of the disjuncture between an Indian reserve geography premised on access to fish and a legal regime that had detached the fisheries from the reserves. It was not something for which the commissioners were prepared, but they quickly discovered that questions of land and fish were inseparable. In community hearings around the province, a great many Native chiefs spoke to the need for more and larger reserves to secure, among other things, access to the fisheries, but they also spoke of non-Native incursions into their fisheries, of their increasing alienation from the commercial fisheries, and of their understanding from earlier reserve commissions that even if their land base were small, the government would protect their rights to the fisheries. In short, in attempting to deal with the land question, the commissioners could not escape "the fish question." This chapter deals with that question through an examination of the Royal Commission on Indian Affairs and its aftermath.

The Royal Commission on Indian Affairs, 1913-16
By 1910, when Indian Superintendent and Reserve Commissioner A.W. Vowell resigned, the process of Indian reserve allotments had ground to a halt in British Columbia. The provincial government, a reluctant participant from the beginning, refused to set aside any more land for reserves. It thought existing reserves too large and claimed a reversionary interest in land removed from the reserves, including a right to the proceeds from its sale, cut timber, and minerals. Wilfrid Laurier's Liberal government in Ottawa, still uneasy about the province's refusal to recognize Native title and not prepared to abandon the Indian land question without at least

concluding the process of reserve allotments, was not in a position to impose its view on the province. As a result, when Vowell retired, the position of Indian Reserve Commissioner disappeared.

The Indian land question, however, did not disappear. In fact, growing and increasingly organized Native protest – including delegations to London and Rome in 1904, London in 1906, and Ottawa in 1908 and 1909 – and the emergence of three associations – the Nisga'a Land Committee in 1907, and the Interior Tribes and the Indian Rights Association in 1909 – had raised its profile and brought it to an international stage.[1] After final attempts to induce provincial participation in a continuing process, the Dominion government prepared a reference to the Supreme Court of Canada on the question of Native title, something the province would not agree to join.[2] However, before the questions were put to the court, the Liberal government fell in the 1911 election. The new Conservative government under Robert Borden declined to press the controversial issue on behalf of a constituency that had no vote, with the prospect of antagonizing one that did. Instead, Prime Minister Borden opened discussions with the Conservative premier of British Columbia, Richard McBride, and then appointed an official from the Department of Indian Affairs (Indian Affairs), James A.J. McKenna, as a special commissioner of the Government of Canada to investigate Native concerns and to negotiate with the province to develop a process for addressing these concerns. The result, in September 1912, was the McKenna-McBride Agreement, which established a joint Royal Commission on Indian Affairs for the Province of British Columbia, a commission that became known as the McKenna-McBride Commission.[3] The agreement was a compromise: the province would concede its claim to the reversionary interest in the reserve land, the Dominion that the question of Native title would be, in McKenna's words, "dropped."[4] But if the question of Native title had been raised and dropped, the fisheries were not even a part of the discussion. In the postscript of a long letter to Premier McBride, McKenna noted: "I have not touched upon the complaint of the Indians as to the regulations restricting hunting and fishing. That awaits investigation, but does not at all stand in the way of the settlement of the land question."[5] McKenna was to be proven wrong almost immediately once the commission began its work.

Under the terms of the agreement, the work of the McKenna-McBride Commission was entrusted to five commissioners – two appointed by the province, two by the Dominion, and a chairperson to be decided upon by these four. The province named James P. Shaw, a member of the provincial legislative assembly from Shuswap, and Day H. Macdowall from Victoria, while the Dominion appointed McKenna and Nathaniel W. White, KC, a

Photo 8.1 Members of the Royal Commission on Indian Affairs for the
Province of British Columbia (McKenna-McBride Commission) on the steps of
the provincial legislature in Victoria, 1913. Front row from the left: Nathaniel
W. White, James A.J. McKenna, Edward L. Wetmore, and James P. Shaw. Back
row: unidentified. | *BC Archives A-08795*

senior lawyer from Nova Scotia. These four then appointed Edward L.
Wetmore, a retired judge from Saskatchewan, as chairperson. He served
for only six months before resigning, and the remaining commissioners
chose White to replace him as chair. The Dominion appointed Saumarez
·Carmichael, KC, a lawyer from Montreal, to fill White's position (Photos
8.1 and 8.2).

These men began their work in May 1913, establishing a base in Victoria
and then setting off on their first field excursion, working their way north
from Victoria up the east coast of Vancouver Island. Their first stop was the
Comiaken village on the Cowichan reserve near the mouth of the Cowichan
River. The Cowichan opened the meeting with a prayer and a hymn. The
commissioners responded by reading their commission. Wetmore then explained that he and his fellow commissioners were to recommend the final
configuration of Indian reserves, that this might mean enlarging or reducing existing reserves, and that the intent was to eliminate the province's
reversionary interest so that the Cowichan could sell their land to raise
capital if that were what they wished. He emphasized that the provincial

Photo 8.2   The royal commission in session, Victoria, November 1913. |
BC Archives H-07035

and Dominion governments had compromised to overcome their differ-
ences and to establish the commission, and he urged the same reasonable-
ness from his audience. They might not get all the land they wished, and,
further, it was not for them to say whether certain reserve reductions were
in their best interests or not; any cut-offs would be more than made up for,
he suggested, by their enhanced interests in what remained. This speech,
which omitted even passing reference to the question of Native title, was a
variation of that which the commissioners would repeat in each Native
community.[6]

   Chief Charlie Suhiltun (Seehaillton) from the Quamichan village was the
first to respond, thanking the commissioners for their visit, invoking the
Cowichan's loyalty to His Majesty King George V, and then turning immediately to what would be the two principal concerns in almost all coastal
communities: insufficient land and lack of access to the fishery. "The reason I speak the best way I can for my people," he said, "is that not only is
the land very small but the white people are making laws to stop them
getting their food from the River."[7] The issue was particularly pressing for
the Cowichan, who were embroiled in a long dispute with the Department
of Marine and Fisheries (Fisheries) over their fish weirs in the Cowichan
River, but the commissioners would hear similar complaints in virtually
every corner of the province.[8] Expecting to deal with land issues, they were
unprepared to answer questions about the fishery, but they promised to
investigate and to respond. They asked the commission's legal counsel,

Photo 8.3 Chief Charlie Suhilton at the meeting with the royal commission, 27 May 1913. Note the Union Jack flying from the carriage, and the cap that was presented to him by Governor James Douglas. Both are displayed as an indication of loyalty to King George V, but also as an appeal to the Crown's promises – from the Royal Proclamation of 1763 to the Douglas Treaties of the 1850s to the Cowichan chiefs' audience with King Edward VII in 1906 – of just treatment and of the rule of law. | *BC Archives H-07043*

McGregor Young, for an opinion on the hunting and fishing regulations affecting Indians. Within a few days of their visit, and before Young could report or the commissioners could respond to the Cowichan, Fisheries officers destroyed the weirs on the Cowichan River. This provoked a stern rebuke from commission chair Wetmore. Would not the raid be seen as a consequence of the commission's visit, he asked, and how could the commissioners gain the trust of Native peoples when Fisheries destroyed the weirs?[9]

Young produced a report suggesting that Indians could fish for food at any time, including close seasons, if they held a food fishing permit from the inspector of Fisheries. If they fished during a designated opening, they could sell their catch.[10] Hoping it might forestall further enquiries and complaints, the commission issued a statement outlining the law "for the information of Indians."[11] However, Young's interpretation that the regulations allowed the sale of fish caught under a food fishing licence was at odds with Fisheries' intent, and to make this clear the department amended the fishing regulations in 1915 in an effort to eliminate the sale of food fish (see Chapter 6).

As they travelled the province over the next three years, the commissioners heard a similar message: Native peoples wanted secure access to their fisheries, and they wanted the government to limit the incursions of non-Native fishers. In a meeting at Rivers Inlet in August 1913, Oweekeno spokesperson Joseph Chamberlain wondered how the canneries had come to occupy their land. "The reason we want to get this land from Quay to Smith's Inlet," he continued, "is to make our food supplies secure, and so that we will be able to keep the fish for ourselves."[12] Later that month at Bella Bella, Heiltsuk chief Moody Humchit began by focusing on the connections between the reserves and the fisheries:

> I would like to say in respect to the Reserves which were set aside for the Bella Bella Indians some time back, that I was with the surveyors at that time, employed by them, and I understood that these Reserves were set aside for the exclusive use of the Indians. I think we ought to enjoy exclusively the hunting, and particularly the fishing privileges, on these Reserves and in the vicinity of these reserves, which we do not enjoy at the present time.
>
> The Chairman: – Do I understand you to say that as these Reserves were set aside for the Indians, the Indians should have exclusive fishing privileges in all the Inlets on the Sea around here?
>
> A. Yes, everywhere. We are more anxious than ever, at the present time to have these things put right. It has been a very poor year with us.[13]

Humchit indicated that they were not asking to fish without licences, but only that they should be able to take up the licences that had been granted to the cannery operators in the region. He was referring to the seine licences held by Draney Fisheries Ltd.[14]

In October 1913, the commissioners returned south to meet with the Okanagan. Farming and grazing land, and water for irrigation in the dry southern interior, were their principal concerns, but voices speaking to the fisheries were persistent. Chief Pierre Michel was one of several who testified about the increasing enforcement of fisheries and game laws against Natives who used traditional fishing technology, primarily spears and weirs. "We were told by the policeman," he said, "that we must not use a spear to catch the fish – We cannot catch them by the naked hand, and I don't think there is one of my people ever went to the river to get any salmon this year – They were all afraid."[15]

Before they embarked on their field work in 1914, and suspecting that they would encounter repeated questioning about the fisheries from the Nuu-chah-nulth and Kwakwaka'wakw peoples on the west and northeastern coasts of Vancouver Island, the commissioners convened a meeting in April with the senior officials from the Dominion departments of Fisheries and Indian Affairs and the provincial Department of Fisheries. Commissioner Macdowall, the acting chair, characterized the meeting as "an informal consideration of fishing questions as they especially affect the Indians, in the hope that some solution thereof may possibly be arrived at which the Indians may be better satisfied, and without sacrifice of the principle of conserving the fish of British Columbia."[16] The conflict over fish weirs on the Cowichan River occupied much of the discussion, and the only firm resolution from the meeting was that government officials would convene another meeting with the interested parties in an attempt to resolve that dispute. Other questions arose, such as Native access to independent licences on the central and northern coasts, and the sale of fish caught under a food fishing licence, but about these there was no agreement. Even the Fisheries officials were divided over Native access to independent licences. The province's deputy commissioner of Fisheries, D.N. McIntyre, and the Dominion's chief inspector of Fisheries for British Columbia, F.H. Cunningham, thought that "the Indian would never be satisfactory as a free fisherman as he was less reliable and would be apt to break contract and leave the employing canners in the lurch."[17] Inspector J.C.T. Williams, who was responsible for District 2 (central and northern coasts), disagreed. "Indians," he argued, "should at least have the opportunity to make good."[18] The commissioners withheld their opinions. Still early in their mandate, it appears they were trying to reduce the conflict over fish and, by doing so, to disentangle the fisheries from their work identifying the appropriate land base. If that were their goal, they were unsuccessful.

In May, the commissioners worked their way through Nuu-chah-nulth territory along the west coast of Vancouver Island. Discussions about reserves invariably turned to concerns about the fisheries, and the testimony from the communities on Barkley Sound and along the Alberni Canal provides a good example. Nuu-chah-nulth witnesses spoke repeatedly about the increasing enforcement of the provisions in the *Fisheries Act* and *Regulations,* particularly those prohibiting fish weirs and traps, and the impact this had on their ability to procure sufficient food. Tseshaht (Sechart) chief Shewish testified that his people had once used fifteen traps on the Somass River at the head of the Alberni Canal, but such weirs were now prohibited. How were they to catch their fish? Indian Reserve Commissioner Peter O'Reilly, who had allotted nine Tseshaht reserves in 1882, had described Tsahaheh Reserve No. 1, a comparatively large reserve on the Somass River, as "their most valued Salmon fishery." Fisheries had closed it, setting aside the portion of the river running through the reserve as a protected spawning ground. The Hupacasath (Opetchisaht) also had reserves on the Somass. Chief Dan Watts told the commissioners that his people understood they had been promised undisturbed access to the river – perhaps he was referring to O'Reilly's representations – but this was not their recent experience:

> Another thing I would like to say about the fishing in this River (the Somass) here. Many years ago the big men told all the Indians they could fish in this river all they wanted for their food, but now these whitepeople try to stop us. I don't know what we are going to do. We live on our fish – we are not like whitepeople – it is hard for us to get a job here the old people cannot get any job from the whitepeople because they won't employ them. When the young people go out fishing, they give it to their friends, and if the whitepeople are going to stop us fishing with gill nets I don't know what we are going to do – and we don't want to be stopped. We always want to fish. They stop our traps up the river there. The purseine [*sic*] does more damage than we do. Our net is only ten feet wide, and they stop us from using that.[19]

Other witnesses wondered why they were not allowed to sell fish even to procure other food such as flour, sugar, and tea. Hupacasath spokesperson Tatoosh made this statement:

> I want to tell you about the salmon fishing on the river. We used to have at one time great fishing on the river and we used to sell our salmon, but when the whitepeople came in here they stopped us from selling the salmon. The whites brought tea, sugar and flour in here and we have to eat it, but where are we going to get the money to buy all those things. We cannot sell the fish, and we cannot work, because the whitepeople won't employ us. We want to know if we have the right to catch and sell fish in season and out of season.

The commission chair, Nathaniel White, responded that he thought they would be much better off if they turned to raising vegetables and cattle. How this might have provided a viable livelihood on the mountainous, heavily forested west coast of Vancouver Island was not clear. Tatoosh continued:

I don't understand why the Government don't allow the Indians to sell fish out of season. It is not the Indians fault. Years ago the fish used to be plentiful, but since the canneries have come here there has been hardly a fish left. Why does not the Government allow the old people to make a few dollars out of the fish. They should not allow purse seines, they ought to give us what we used to live on before instead of gobbling the whole thing. Another thing there is a dam up here. Since it has been there hardly any fish has passed there, and spring salmon and co-hoes never go past that dam. The Indians never did that, it is the whitepeople that is decreasing the fish.[20]

Tseshaht spokesperson Mr. Bill added his concerns:

I am very glad to meet you people – just like meeting my father. My only complaint is about the fishing. I don't know why they won't allow us to catch such few salmon – 4 or 10 – We have to buy sugar, tea and flour and all such stuff as that, and I can't eat unless they allow us to catch the fish.

I would be very glad if you would give us the privilege of selling fish – I don't know why the whitepeople won't allow us to fish on their river. We have claimed this river ever since the Indians were made. The Government or the Fishery Inspectors did not bring the fish into the rivers after they came into the country – The salmon were here even before we were – We claim the salmon ourselves and it should not have anything to do with the whites – We were here before they were and we claim all the fish. I am not doing any harm to anything that the whites brought to this land, such as pigs, hens and horses, and we only use what was given to us by Jesus for to eat and to use and to make our living on. When this earth was made and this river with salmon in it and the forests with deer in it and all that we use, they were made for us to use and everything that was in it, and they used to have wars (the Indians) with the other Tribes and we got this river, running up here and our title with it. As the Chief said about the purse seine, I don't know why they don't allow us to fish with traps but they allow all the big firms to fish with seines and they kill all the fish, and very soon we won't have anything. I am very sorry the Government allows the use of seines to come in here and kill all the fish.[21]

Tatoosh and Bill, along with numerous other Native witnesses and the Indian agent for the West Coast Agency, Mr. Cox, testified that the cannery-operated seine nets were depleting salmon runs in many of the small rivers along the west coast, including those bounded by reserves. In Barkley Sound and the Alberni Canal, Fisheries had issued three drag- and two purse-seine licences to Wallace Fisheries Ltd. (Figure 8.1), and the company fished these waters aggressively. Uchuklesaht chief Charlie Jackson testified that the company had even placed a donkey engine on the Uchuklesaht reserve, Elhlateese No. 2, to pull its drag seine ashore. It had done so without permission or compensation, and company workers had intimidated Uchuklesaht fishers. Jackson thought the Uchuklesaht could coexist with Wallace Fisheries so long as the company paid for the privilege of using the reserve and its foreshore and did not take all the fish.[22]

In December 1915, their field work almost complete, the commissioners met again with senior Dominion and provincial Fisheries officials.[23] They had recently returned from hearings with the Nisga'a and Tsimshian, where fishing issues were to the fore, but fishing rights had been a concern everywhere. Unlike earlier meetings with Fisheries officials, which had been exploratory in nature, this time the commissioners asked hard questions. They were the questions that Native peoples had been asking them, and to ensure that none were missed, the commissioners had all the Native testimony involving fisheries extracted from the records of two and a half years of hearings, compiled in one volume, and indexed in preparation for the meeting.[24] At its outset, commission chair Nathaniel White repeated three of the central Native concerns over the fisheries: first, the inequity of a policy that denied Native access to independent licences on the central and northern coasts; second, the restrictions on food fishing licences that prevented the sale of fresh fish even to purchase other food; and third, the status of exclusive fisheries in waters running through Indian reserves.

The first issue, Native access to the commercial fishery, produced an extended and heated discussion as the commissioners challenged the Fisheries officials to justify their policy. The following excerpt provides a sense of the content and tone of the meeting. After eliciting an admission of the department's preferential treatment of white fishers, Commissioner McKenna asked:

> Q. And you discriminate against the Indians who were the first settlers on the Coast, who built the first boats and knew the harbours and knew the places where the fish ran, and these Indians cannot have the same privileges as these white men because they are bronze skinned?
> Mr. Babcock: The fishing grounds on the Skeena River are not what they were ten years ago.
> Mr. Commissioner McKenna: You would not expect them to be.

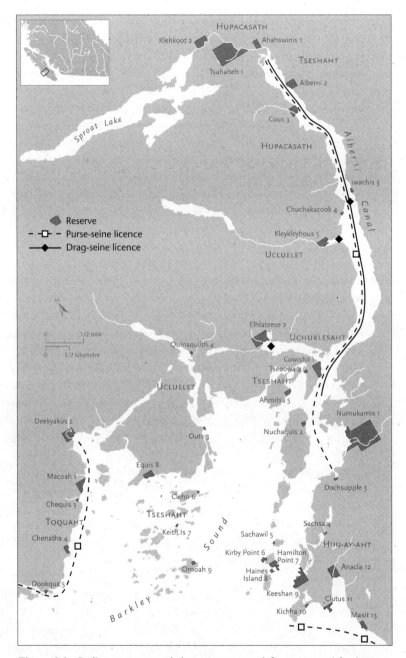

Figure 8.1 Indian reserves and the areas reserved for commercial seine licences in Barkley Sound and the Alberni Canada, 1915-16. Wallace Fisheries Ltd., which held the seine licences, had placed a donkey engine on the Uchuklesaht reserve Elhlateese 2 to haul drag seines ashore without securing the permission of the Uchuklesaht.

Mr. Commissioner Macdowall: You say that you give these independent licenses to assist the white men who go in to settle up that part of the country – that is a very good thing; but in your treatment of the Indians you are depriving him of his means of making a livelihood. He has been fishing up there from time immemorial. By what means of justice or right or right-dealing towards these aborigines – by what law are you working under that you are authorized to deprive him of his only means of making a living – by what law do you do that?

Mr. McIntyre: We would challenge your first statement: that is the Indians who have made their livelihood on the Skeena River. The Indians on the Skeena River were very scarce even before the white men ever went there – even the Indians who fish on Rivers Inlet are not natives of that District ... If the canneries were not there the Indians would reap no benefit at all.[25]

A few exchanges later, the following question and response:

Mr. Commissioner McKenna: Suppose I am an Indian – I am on the Skeena River, and I am cultivating a piece of land on the Skeena River in the North; I have a boat and a little money. I can go and get credit; I can go and get a whole outfit, and I am doing the best I can and remember I am a British subject, and I might say that these Indians in the year 1885 produce the most revenue for the Province of British Columbia. Now remember that I am an Indian and I go and I ask for an independent license; I am refused but Tom Jones comes along and clears his piece of land, and the first application that he makes for an independent license it is granted to him without any question at all. Now can anyone tell me why I am discriminated against when I have the boat and everything to fish with?

Mr. Cunningham: Is it not a fact Doctor that the Indian is a ward of the Government?

Mr. Commissioner McKenna: Not at all.

Mr. Cunningham: If he is not taken care of why do they have Indian Agents to look after them; why do they give them farming implements and cattle and in fact everything they want?[26]

These views of Native people were widely held. McIntyre's opinion, echoing that of others in the industry (see canner Henry Bell-Irving's comments in Chapter 7), was that Native peoples were an ephemeral presence in the province until the "white man" appeared, and it was only then that their productive capacity could be fully realized as wage labourers. It was as though the white man and his capital brought the Native presence into being. Their previous existence, as their previous use of the resource, was fleeting and certainly not a valid foundation for claims of Native title or fishing rights. To the extent that Native people did have a claim against the Crown, it was a moral claim that the Crown more than adequately fulfilled, Cunningham argued, through Indian Affairs and the Indian agents. Native peoples were wards of the state. As such, they were well looked after and had no claim to the fisheries.

Following this meeting, the commissioners produced a report on the fisheries. It included a short memorandum on the "Fishing Rights and Privileges of Indians in B.C.," accompanied by numerous appendices of commission transcripts, Native petitions, government letters, and other documents dealing with the fisheries. Presaging their final views, the commissioners denounced the whites-only policy for independent licences on the north coast: "The Commission is unanimously of the opinion that the Indians of Northern British Columbia are – but should not be – discriminated against in the issuance and use of these 'independent' fishing licenses; and that there is no authority conferred by the Act, or intent therein expressed or suggested, for such class or racial discrimination." They urged Fisheries to follow through on its policy to require the canneries to hire Natives to work the cannery licences. The commissioners also thought that the validity of earlier fishing rights granted by reserve commissioners should be investigated and that canneries should only be allowed to use reserve foreshores with the consent of, and compensation paid to, the Native inhabitants. Finally, they advocated for a system of "peddlers' licences" to allow the local sale of fresh fish.[27]

The McKenna-McBride Commission issued its final report in June 1916.[28] It contained a mass of information, organized by Indian agency and displayed in tables, including details of existing reserves and the people who lived on them. These tables, Cole Harris has argued, were the commissioners' way of reducing the volumes of Native testimony, the surveys, the census data, and the commissioners' views into a manageable and quantifiable format. The numbers and the highly abbreviated, standardized comments about the nature of the reserves and the lives of the people who inhabited them allowed the commissioners to draw comparisons and adjust the land base of the communities to fit a relatively standard pattern.[29]

There had been 1,085 reserves when the commissioners began, covering 753,708 acres of British Columbia.[30] Of these, 455 (42 percent) were described in the final report as fishing stations or as reserves where fish were one of the principal products of those who lived on the reserve. Most of these identified fishing stations overlapped with the reserves that earlier commissioners had specified as fishing reserves in the initial grant. There were a great many more reserves where the commissioners listed fishing as a "chief occupation" of the inhabitants. In fact, there were very few reserves where fishing was not included among the chief occupations, usually with some combination of trapping, hunting, farming, ranching, and various forms of wage work, including cannery work. To this land base, the commissioners proposed an additional 484 reserves and fifty-five reductions or cut-offs. The acreage of the new reserves was almost twice that of the reductions and cut-offs, but the value of the new land was estimated to be only one-third of that which would be removed. In the minutes of decision creating the reserves, the commissioners described 142 of the new reserves (29 percent) as fishing stations. Most of these were on the north coast in Tsimshian and Nisga'a territory, and in the northern interior, territory of the Dakelh, Sekani, and Tahltan.

The additional fishing stations notwithstanding, it is clear from the report that the commissioners did not know how best to protect Native fisheries or whether they had a mandate to do so. They appear to have been unsure about the status of the earlier grants, particularly O'Reilly's grants of exclusive fisheries (described in Chapter 3), and of their power to confirm them. In one of their last hearings, the Indian agent for the New Westminster District, Peter Byrne, described how Fisheries officials shut the Mount Currie (Pemberton) fishery in Lillooet Lake and its feeder rivers and creeks to conserve fish for a hatchery. Commissioner McKenna questioned Byrne about the exclusive right to fish on the Lillooet River that O'Reilly had reserved to the Mount Currie band and whether this had any impact on Fisheries officials. Byrne indicated that he had raised the issue with senior Fisheries officials. Nothing had come of that meeting, but he thought that the exclusive fishery should be recognized as part of a treaty between the Crown and the Mount Currie band.[31]

In 1916, Commissioner Macdowall wrote to the deputy superintendent general of Indian Affairs, Duncan Campbell Scott, to ask for his opinion on whether O'Reilly had acted within his authority when allocating exclusive fisheries. Scott replied that he did not think O'Reilly had the authority to make such grants, a position that Indian Affairs had conceded to Fisheries in the late 1890s, but added that he hoped the commissioners would confirm the grants.[32] Macdowall, who, as one of the provincial representatives, was not sympathetic to the idea of increasing reserves in size or number, replied that he was part of a majority on the commission (including McKenna and Shaw) that would support a motion to confirm, so far as they could, "the Indians in their fishing privileges."[33] In the final report of the royal commission, at the end of the section for every agency where O'Reilly had allocated fishing rights, the commissioners included the following resolution:

WHEREAS former Indian Reserve Commissioners, acting under joint Governmental Agreements, allotted defined Fishery Rights to certain Tribes or Bands of Indians in British Columbia;

WHEREAS this Commission has been unable to obtain any advice from the law officers of the Crown in right of the Dominion of Canada as to the authority of the said former Commissioners to allot such fishery rights;

AND WHEREAS this Commission desires that any right or title which Indians may have to such allotted fisheries may not be adversely affected by inaction on its part —

BE IT RESOLVED: That, to the extent to which the allotting Commissioners had authority to allot such Fishery Rights, this Commission, insofar as the power may lie in it so to do, CONFIRMS the said allotted Fishery Rights as set forth in the Schedule hereto appended.[34]

The appended schedule included the text of O'Reilly's minutes of decision for each reserve in the agency that explicitly included fishing rights. These resolutions reflected the commissioners' uncertainty about O'Reilly's jurisdiction and their own, but they also reflected the commissioners' determination to do what they could to secure Native fishing rights.

In addition to the official report, the commissioners authored a twenty-eight-page "Confidential Report."[35] It contains material that, it appears, the commissioners thought too inflammatory and destabilizing for public consumption. One-third of that report dealt with the fisheries, and in it the commissioners made many recommendations. Regarding Native access to commercial fishing licences, the commissioners were astonished and dismayed that Native fishers could not acquire independent licences on the central and northern coasts. There was no foundation in law for such a policy, they wrote, and it should end immediately. Native peoples must be treated as other British subjects. Concerning the "attached" or cannery-held fishing licences for the northern and central coasts, the commissioners reported that although Fisheries required the canners to hire Native fishers first, this was not happening. Instead, canners were hiring Japanese fishers, shutting many Native fishers out of the industry. Furthermore, Fisheries had embarked on a strategy to eliminate the cannery-held licences, replacing them with independent licences. This policy, combined with that limiting independent licences to white fishers, would lead to the complete exclusion of Native fishers, something so appalling that the commissioners could not believe that the larger government was aware of Fisheries' policy: "The Commission feels, however, that the Government of Canada cannot be aware that under one of its Departments a policy has been designed and is being enforced which will exclude the Indians of northern British Columbia from the salmon fishing industry."[36] The commissioners also recommended that Native peoples should be encouraged and assisted to participate in the fishing industry not just as fishers and cannery workers, but as processors and cannery operators. Their full participation in the industry, the commissioners wrote, would benefit everyone.

The commissioners then addressed the "special fishing privileges," including the exclusive fisheries and the designation of many reserves as "fishing stations." They understood these privileges as Crown grants and, referring to the debate between Fisheries and Indian Affairs in the 1880s, were uncertain whether their predecessors in the Indian Reserve Commission had had the legal authority to make such grants. If the legal authority did not exist, the commissioners suggested that steps should be taken to convert what was a "moral right" to the named fisheries into a "legal right."[37] Similarly, where rivers flowed through reserves, the exclusive right to fish in those waters should be reserved to the Indian band through whose reserve the river flowed, if this had not already been done. These rivers were only to be open to the public if the band gave its "formal consent" and received "due compensation."

The disjuncture between the purpose for which a great many of the reserves had been allotted in British Columbia and the Natives' access to the fisheries, greatly hindered by the seining licences that blanketed the central and northern coast, was not lost on the commissioners. They were insufficiently bold to insist on the transfer of licences or a similar remedy, but they were clearly concerned that other people were working these fisheries without the consent of, or compensation for, Native peoples. They recommended that the connection between reserves and fisheries "should be made the subject of careful inquiry so that the purpose for which the fishing stations were established be preserved."[38]

Finally, in response to the many complaints about the prosecution of Native fishers who sold a few fish in local markets without a commercial licence, the commissioners recommended the creation of a peddler's licence that would allow the sale of fish so that other food might be purchased. This, the commissioners hoped, would reduce the calls on Indian Affairs for assistance.

The commissioners had done what they thought they could. Perhaps it had some limited effect. In 1919, Fisheries allowed Native fishers to apply for independent licences on the north and central coasts. Seine-net licences were still only available to white fish processors, primarily cannery operators, and remained so until the 1924 season. But if the formal barriers to Native fishers in the industrial commercial fishery gradually disappeared, many informal barriers, including lack of access to credit and pervasive discrimination, remained. In the food fishery, restrictions and levels of enforcement only increased in the decade following the McKenna-McBride Commission's report.[39]

From Report to Resolution, 1916-24

Even with the most pointed critique of their policies tucked safely away in the confidential report, the Dominion and province did not release the royal commission's final report to the public until 1919. The only activity in the interim had been mounting Native protest over the lack of progress on the many issues surrounding reserves (size and location, non-Native incursions, the province's claim to the reversionary interest, and ownership of the foreshore), the restricted access to food and commercial fisheries, and the continuing failure to address Native title.[40] Recognizing that something needed to happen, in 1919 the provincial legislature passed the *Indian Affairs Settlement Act,* which authorized the government to conduct the negotiations necessary to implement the report.[41] The Dominion moved more slowly, unsure what to do about the provisions in the *Indian Act* that required Native consent before reserves could be reduced or cut off.[42] Officials knew it was highly unlikely Natives would give this consent, but they were also aware that the BC government would insist on the cut-offs and reductions suggested in the report if British Columbia were ever to accept it. To circumvent the need for Native consent, the Dominion government passed the *British Columbia Indian Lands Settlement Act* in 1920. The act overrode the provisions in the *Indian Act* and gave the government authority to do what it had promised it would not: agree to reserve reductions and cut-offs without the consent of Native people.[43] At this point, the Dominion and the province began negotiating the agreements that would lead to the implementation of a slightly modified version of the royal commission's final report.

At the province's suggestion, the two levels of government appointed representatives to a two-person panel that would review the report and make final recommendations. The province nominated Major J.W. Clark; the Dominion, W.E. Ditchburn, chief inspector of Indian Affairs for British Columbia. Although it had discounted the need for Native consent, Indian Affairs employed James A. Teit, ethnographer and advisor to the Nlha7kapmx and former secretary of the Interior Tribes, to provide a perspective on the views of Native peoples.[44] Teit was incapacitated by, and then died from, cancer shortly after his appointment, and Indian Affairs appointed Peter Kelly (Haida), Andrew Paull (Squamish), and Ambrose Reid (Tsimshian) to replace him. Kelly and Reid were the president and secretary of the Allied Indian Tribes of British Columbia, a coalition of tribal groups that had formed in 1916, replacing the Indian Rights Association, to protest the lack of recognition of Native title in British Columbia. As advisors, however, their input was restricted to the placement of reserves; neither Dominion nor province would talk about Native title.

Ditchburn still hoped to secure some form of Native consent to an amended version of the royal commission's report, and he thought initially that the Allied Tribes might be the vehicle through which to achieve it. However, he eventually concluded, along with Superintendent General for Indian Affairs Duncan Campbell Scott, that although the Allied Tribes had broad support from coastal groups, it was insufficiently representative, particularly among interior tribes. Perhaps more to the point, the senior Dominion officials could not accede to the conditions for settlement proposed by the Allied Tribes, and they knew that provincial officials were even less flexible. As principal spokesperson for the Allied Tribes, Kelly pointed to the Crown's continuing failure to recognize Native title and the increasingly restricted access to the fisheries as the key grievances. Native fishers still had no access to seine licences; cannery seines continued to fish in waters adjacent to reserves and sometimes established beachheads on the reserves themselves; fewer Natives held independent fishing licences; and Fisheries officials seemed to be closing down the Indian food fisheries. The Allied Tribes would not agree to the McKenna-McBride report, however modified, fearing that acceptance would compromise Native title claims and fishing rights. In 1922, Ditchburn reported his understanding of the conditions surrounding the fisheries that would have to be included in any comprehensive settlement that Native leaders might accept:

> From what I have been able to gather the Indians will insist that there be no interference for the future in connection with their being able to obtain their supply of fish for food purposes. The Coast Indians will also ask to be given absolute fishing privileges for commercial purposes in waters fronting on and running through their reserves, but I cannot see how they can hope for success with regard to streams as they would deplete the supply at the spawning grounds, but the setting aside of fishing areas fronting reserves is worth consideration. I feel, however, that they might be assured of being able to earn a livelihood at fishing without having to obtain licenses to use nets. At present Indians can purchase a gill-net license but are discriminated against by the Fisheries Department in the matter of seining licenses, and many of them are in a position to own a sein [sic] ...

In the setting aside of reserves for the Indians on the Coast the per capita acreage
has been very small as it was always considered that they would be enabled to earn
their living by fishing and hunting and it would be unfortunate if any impediment
were placed in the way of their being able to do so.[45]

Fulfilling their advisory roles, Kelly, Paull, and Reid provided Ditchburn
with a long list of suggested new reserves, many of them fishing sites.
Ditchburn thought the list highly unsatisfactory, and he reduced it considerably, removing all parcels that had already been alienated or which he
thought excessive.[46] The provincial representative, J.W. Clark, had little
use for even this reduced list. In his opinion, "the choicest building sites
for industrial development at the mouths of the best rivers in the Coastal
District are already covered by existing Indian Reserves."[47] He was more
interested in reducing existing reserves than in adding to their number. In
the end, Ditchburn and Clark submitted separate but substantially similar
reports that, with some minor tinkering, largely confirmed the royal commission report. Ditchburn's efforts to increase the size and number of reserves and Clark's efforts to reduce them largely cancelled each other out.[48]
This was a result that satisfied the province. In July 1923, it accepted the
McKenna-McBride Commission's report as modified by Ditchburn and
Clark.[49]

Again, the Dominion hesitated. Officials in Indian Affairs still hoped
that Native leaders might be convinced to accept the reports, and they
convened meetings with Native leaders in July and August 1923. The fundamental disagreement over Native title remained the insurmountable obstacle. Native leaders insisted that Native title must be part of any discussion;
officials within Indian Affairs demurred, uncertain about the legal status of
Native title and knowing they would never reach an agreement with the
province if they admitted its existence. The other intractable issue was access to the fisheries, and much of an August meeting between senior officials from Indian Affairs, the executive of the Allied Tribes, and other Native
leaders revolved around fishing rights. Just as Native leaders refused to
separate the question of title from the allotment of reserves, so they refused to separate the question of access to the fisheries. If they were to
agree to the report, certain fishing rights had to be guaranteed.[50]

"The fishing question," said Peter Kelly at the outset of a long discussion
over two days, "is a burning one, and it does not just affect one particular
section of the Province, but it applies to the Province as a whole."[51] He and
others recounted many instances where Fisheries officers had destroyed
fishing gear (weirs and nets), confiscated fish, and levied fines against

Native fishers. The lack of access to seine and independent licences was a central concern, and so was the failure to recognize exclusive fisheries associated with reserves. Kelly pointed to the Metlakatla community that had moved from British Columbia to Alaska in the 1880s and whose right to an exclusive fishery stemming from their reserve grant had been confirmed in 1918 by the United States Supreme Court in *Alaska Pacific Fisheries v. United States*.[52] Why, he wondered, would the Canadian government not do the same?

At the beginning of the second day, Kelly summarized the position of the Allied Tribes. Under the rubric of fishing for domestic purposes, he argued for a right to fish for food anywhere in British Columbia, including tidal waters, without a permit or limits as to quantity. As for commercial fishing, Kelly claimed first that Natives should have a right to troll for salmon without licence in tidal waters and sell the fish to whomever they wished. Second, Native fishers should have the right, hitherto denied, to acquire a seine licence, and the licence fee should be cut in half. Beyond that, Native fishers should have the same rights as any other citizen to fish in any waters. Third, Natives should have the exclusive right to fish with seine licences on Indian reserve foreshores and at the mouths of rivers that run through reserves. Fourth, certain fishing areas should be set aside for the exclusive use of Native fishers, much as the waters surrounding the Annette Islands had been set aside for the Metlakatla in Alaska. Finally, although Native fishers were now eligible for independent drift-net licences, Kelly sought to ensure that any residual discrimination would not affect their access to these licences.[53] Regarding the connected issue of the foreshore, Kelly insisted that the Crown recognize Native title to the low-water mark in front of reserves. If Natives were to accept the report of the royal commission – in effect concluding a treaty in which they would surrender their Native title – these were the minimum requirements with regards to the fisheries.[54]

Realizing that there was no middle ground between Native leaders and the provincial government on the question of Native title, or between Native leaders and the Department of Fisheries on the question of fish, in 1924 the Dominion moved to accept the royal commission's report as modified by Ditchburn and Clark.[55] Although fisheries issues had been at the fore throughout the two-year process to reconsider reserve allotments, the only tangible results were a handful of new coastal reserves, three of which were recognized as fishing stations (two in Kwakwaka'wakw and one in Tsimshian territory), and, in 1924, the removal of the restrictions on Native access to seine licences.

Conclusion

The process that began in 1876 with the Joint Indian Reserve Commission seemed finally to be coming to a close, and although Native leaders were dismayed, officials within Indian Affairs thought they had secured the best arrangement possible with the provincial government. There were still some unresolved details, such as the province's reversionary interest in the railway belt, and there would be a special joint meeting of the Senate and House of Commons in March and April 1927 to inquire into the claims of the Allied Tribes.[56] This was the last time the Dominion government would suffer Native title arguments in an official forum until the 1970s. Shortly after the hearings, it amended the *Indian Act* to forbid the raising of funds to pursue Native title or land claims without the permission of Indian Affairs.[57] The prohibition was repealed in 1951, but it was only after the Supreme Court of Canada's 1973 decision in *Calder v. British Columbia* that the federal government recognized Native title as a legal interest that had to be reckoned with in British Columbia.[58] The Supreme Court would be the catalyst for changes in the fisheries as well, although that had to wait until 1990 and the court's recognition, in *R. v. Sparrow,* of an Aboriginal right to a food, social, and ceremonial fishery – a right that conferred priority over other users.[59]

In 1930, when the Dominion transferred its title in the railway belt and the Peace River block to the province, returning the land that it had received as the province's contribution to the cost of building the Canadian Pacific Railway, it retained title to the Indian reserves in these lands. It was not until 1938 that the province conceded that the Dominion held title to the remaining reserves in British Columbia.[60] At that point the Dominion held 1,536 reserves, amounting to 835,339 acres of land, in trust for Native peoples. This amounted to slightly more than one-third of one percent of the land area in British Columbia.[61] Approximately half of these reserves, that is, nearly 750, had been allotted specifically to secure access to fisheries.[62]

At least so far as governments were concerned, the "Indian land question" was now a closed file. The reserve geography for the province had been set, and for the most part it reflected the province's view. If Native title existed at all, it was a moral obligation that could be met with reserve allotments. Those reserves, moreover, could be small because the economy of most Native communities revolved around the fisheries, including subsistence fisheries and piecework on fish boats and in the canneries. The fisheries were thus inseparable from the land question, but because fisheries were a Dominion responsibility, the debates over access pitted the

Dominion departments of Indian Affairs and Fisheries against each other instead of Ottawa against Victoria. As the province succeeded in denying Native title and restricting reserve size, so Fisheries was successful in denying the recognition of Native fishing rights and confining the food fisheries. These debates revolved around Native peoples and their rights, but Native peoples had no standing except as witnesses appearing before government-established commissions that had little interest in hearing, and even less authority to act on, Native arguments about title and fishing rights.

Conclusion

British Columbia's Indian reserve geography presumes that Native peoples have access to their fisheries. This is revealed numerically: the Indian reserve commissions linked nearly half of the more than 1,500 reserves allotted between 1849 and 1925 to the fisheries. It is displayed in maps that show the proximity of the vast majority of reserves on the coast and in the interior to fish-bearing bodies of water. It is expressed in governmental correspondence in which the Dominion and province justified the land policy in British Columbia, characterized by small reserves and per capita acreages, with the argument that Native peoples in British Columbia were fishing peoples and needed only a small land base to protect their traditional economies and to facilitate their participation in the wage economy. It is underscored by the work of the Indian reserve commissioners, who included exclusive fishing rights in reserve allotments to some bands, and who consistently told Native peoples that their fisheries were secure and their access to them undiminished. It is revealed in the voices of Native peoples, who, to the extent that they were participants in the process, insisted that the governments reserve and protect their fisheries. It is also confirmed in the Douglas Treaties, where the provision for small reserves – "village sites and enclosed fields" – was accompanied by expansive protection for Native fisheries – the right to their fisheries "as formerly." In sum, the reserve geography of British Columbia was built around Native peoples' access to their fisheries. Given the small land base, fish were the one resource that might have enabled Native peoples to build viable reserve-based economies.

When British Columbia joined the Canadian confederation, jurisdiction over Indians and fisheries devolved to the Dominion. Ownership of land vested with the province. Recognizing that the Indian land question had not been resolved in the colony, the drafters of the terms of union included a clause that required British Columbia to provide the Dominion with as much land for Indian reserves as the colony had been in the practice of setting aside.[1] However, past practice had been far from generous and was marked by the colony's refusal, after the Douglas Treaties, to negotiate treaties or to recognize Native title. As a result, the Dominion put itself in the incongruous position of concluding treaties elsewhere in Canada while acquiescing in British Columbia to the province's refusal to negotiate treaties. The making of Native space in British Columbia, as Cole Harris suggests, was driven largely by the province's vision of Indian land policy.[2] Consequently, the Indian reserve geography of British Columbia became a scatter of small reserves, many of which were later cut off or reduced. All but a tiny fraction of the province lay open to immigrant settlement and development.

In describing Indian land policy in the United States, legal historian Stuart Banner suggests that "a reservation could be a prison, if the lock was on the outside, or it could be a haven, if the lock was on the inside. Everything depended on what the reservation was supposed to accomplish." He argues that "by the 1870s, it was obvious the goal of confining the Indians was winning out over the goal of protecting them."[3] This prison/haven dichotomy is too stark when thinking about the role of Indian reserves in British Columbia. The pass system, employed elsewhere to monitor and regulate Native peoples' movement on and off reserves, was not a feature of the reserve system in British Columbia.[4] Indeed, such a system could hardly work where reserves were small and scattered and where frequent movement between them was often essential. As a result, the reserve system in British Columbia did not confine individual Indians. But if individuals were not to be confined, the reserve system was designed to contain the Native presence, opening the rest of the province to non-Native settlement and unencumbered use. The reserves themselves became places of increasingly limited opportunity.

Although the Dominion was a reluctant follower of provincial land policy in British Columbia, it implemented a similar approach in the fisheries, where its jurisdiction was unimpeded. Just as the province denied Native title to the land, so the Dominion denied Native rights to the fisheries. In place of a broad recognition of Native fishing rights or an acknowledgement of the rights reserved in the Douglas Treaties, the Dominion constructed the Indian food fishery as a privilege that performed the same function for the fisheries as Indian reserves did with respect to land. It set aside a small portion of the fisheries for Native peoples, at the same time opening the rest of the resource to an immigrant society.

As a privilege, the meagre allocation of food fish was less secure than the allotted Indian reserves, notwithstanding the many depredations of reserved land. Part of the explanation for this difference lies in the legal regimes that surrounded land and fisheries. Held in trust by the Dominion government for the band to which the lands had been allotted, Indian reserves were unique islands of property within the Canadian state. They were, nonetheless, situated within a regime of private property, and the right to exclude others, which lay at the foundation of that regime, was a right the Indian bands held with respect to their reserves. Although susceptible to competing claims for the land (and the record reveals that Indian reserves were particularly susceptible), those claims had to overcome the general and powerful aversion in land law to interfering with the rights of the private landowner. The regime of private property provided some protection for Indian reserves, even as it was part of a legal apparatus that fuelled the resettlement and development of land outside the reserves.

The ideas of ownership and of private property dominated the understanding of land, but not fish. Native control of the fisheries was unchallenged in British Columbia until the mid-1870s, when, with the introduction of canning technology, an Anglo-Canadian settler society's attention turned towards salmon. Capital flooded into the industry in the last quarter of the nineteenth century, and levels of exploitation rose dramatically. In an era of presumed abundance, a legal regime that loosed the fisheries from their prior moorings in Native custom and tradition and reconstructed them as common property was sufficient to open the fisheries to that capital and to other elements within an immigrant society. As a result, the regime of private property and its right to exclude, deployed on the land, was not, for the most part, reproduced in the industrial commercial fisheries. Instead, the Department of Fisheries imposed a right not to be excluded, and it used this right, embedded in both international and domestic legal regimes governing fisheries,[5] to deny the prior rights of Native peoples to the fisheries. In doing so, the regime of common property became the mechanism of dispossession.

Irene Spry, in her study of the Canadian prairie, makes a similar point. Canada, she argues, displaced the patterns of Native peoples' management, sustained by vast local knowledge, with a legal regime that opened the resources of the prairie to all. Principal among these resources was the buffalo, which, like the salmon of the Pacific coast, was a migratory animal that could not be effectively contained within small parcels of privately held land. A right not to be excluded secured their availability to an immigrant society. The result was the rapid depletion and disappearance of the buffalo, "a classic instance," suggests Spry, "of the 'tragedy of the commons.'"[6] A regime of private property, which further excluded Native peoples from their traditional territories, followed, but the processes of dispossession began with a legal regime that opened the territory and resources to outsiders.

Opening territory that immigrants could occupy and legally possess was the principal role of law – indeed, the principal role of the colonial state – in settler societies. In doing so, the state drew on the repertoire of legal forms that had developed, over centuries, in the home country. It could also turn to legal developments in other colonies, and to this end the Colonial Office in London functioned as a clearinghouse for colonial innovation, assisting in the dissemination of legal forms and other modes of colonial rule around the empire.

The body of the common law that established legal relations between people with respect to things – property law – identified two principal avenues for distributing interests: private property and common property. By the nineteenth century, their use had been largely compartmentalized: private property would spread over land; common property would dominate territory covered by water, including the fisheries. I have argued that, in British Columbia, the intended effect in both cases was to impose a legal regime that dispossessed Native peoples. This need not have been the case. A regime of common property in the fisheries might have co-existed with the recognition of prior Native rights to the fisheries, just as there could have been space for private property alongside Native title. This did not occur in British Columbia, where the province refused to recognize Native title, and the Dominion dismissed Native rights to fish. In this corner of empire, the body of law governing relations between Natives and newcomers, established over several centuries of interaction in North America and overseas, was all but ignored.[7]

Common property was the basic legal template for the regulation of fisheries in the late nineteenth and early twentieth centuries. Private property played a peripheral role, operating over smaller, non-tidal bodies of water. As a result, the right not to be excluded provided the legal basis for those elements in a settler society that sought access to the fisheries. But the right not to be excluded was applied unevenly. Native peoples were excluded from the fisheries where the desire to encourage white settlement or capital investment intervened. On the central and northern coasts, only white fishers were eligible for independent licences, and only white canners or fish processors could secure seine-net licences until the 1920s. These were blatant transgressions of the right not to be excluded, and they were condemned by numerous people in official positions. The right of Native peoples not be excluded was eventually restored, but the effects of the earlier exclusion and of pervasive discrimination hindered Native efforts to re-enter the commercial fisheries. So did the continuing failure to recognize Native or treaty rights to the fisheries. In non-tidal waters, where the law did allow the Crown to exercise its prerogative to grant exclusive fisheries, officials in the Department of Fisheries claimed otherwise. They asserted an erroneous interpretation of the law, but their interpretation, asserted from a position of power, halted the efforts of the Indian reserve commissioners to set aside exclusive Native fisheries.

The common-law doctrine established the basic terrain on which the regulation of the fisheries unfolded. Knowing the lay of that terrain is essential to understanding the regulation of the fisheries, but it is not sufficient. One must go beyond the doctrine to see how the terrain was occupied. In the British Columbia fisheries, the law was applied where useful and ignored where it was thought to be a hindrance or irrelevant, but either tactic almost always worked in the interests of a settler society. Not all immigrants fared equally well, of course. In the fisheries, Japanese fishers and Chinese cannery workers were particular targets of discrimination and of efforts to exclude. Moreover, fishers from all backgrounds often struggled against the collusion of cannery owners to create space for viable livelihoods in a difficult and frequently dangerous industry. Elements within immigrant society benefited disproportionately from the legal regime newly imposed over a territory, but whatever their station, the preponderance of opportunity lay with the newcomers. Such was the effect of the imposition of state law where the power of colonizer and colonized was radically unequal.

This is a decidedly instrumentalist view of the law, not as a unitary and coherent body of independent principles, but as a malleable set of ideas that, in the colonial setting, was used as needed to fortify the position of the colonial state and enhance the interests of a settler society. However, the emancipatory possibilities of law, even in circumstances of great inequality, is one of the recurring themes in the law and colonialism literature. The resistance of indigenous peoples, although unable to stop the imposition of European legal regimes, could in some instances minimize or modify their effects with appeals to the principles embedded within the law or to its procedural detail. Moreover, the colonial state and its laws were not monolithic, as much as they may appear so from a distance. Instead, they were a cacophony of voices often working at cross-purposes, contesting the meaning of English common-law doctrine in its transplanted environment and debating the content and effect of statutory regimes. This is abundantly clear in the regulation of the fisheries in British Columbia.

Of colonialism more generally, anthropologist John Comaroff suggests that, "far from being a crushingly overdetermined, monolithic historical force, colonialism was often an underdetermined, chaotic business, less a matter of the sure hand of oppression – though colonialisms have often been highly oppressive, nakedly violent, unceasingly exploitative – than of the disarticulated, semicoherent, inefficient strivings for modes of rule that might work in unfamiliar, intermittently hostile places a long way from home."[8] By 1925, many immigrant British Columbians hardly considered themselves "a long way from home." They had come to settle in a new home and make it theirs, and they expected the law of the colonial state to help. It did, but it also provided, and provides, spaces – physical spaces, such as the courtroom, or rhetorical spaces of argument – that were, and are, used as sites of contestation and resistance.

Early in the afternoon of Monday, 6 October 1986, two federal Department of Fisheries and Oceans (Fisheries) officers flew a helicopter patrol up the Squamish River. From the air they spotted nets in the river alongside the Squamish Indian Reserve No. 11 Cheakamus (see Figure 0.1 in the Introduction). Several hours later they returned to the river by truck and seized the nets.[9] The next day, fifty-eight-year-old Jacob Kenneth Lewis, a member of the Squamish Indian band, contacted Fisheries to recover his nets.

Lewis held an Indian food fish licence that read: "Jake Lewis ... being an Indian, is hereby licenced to fish for salmon that may be taken in sufficient quantities to be used for the sole purpose of obtaining food for that Indian and his family in the following described waters or area: Squamish River."[10] Lewis and the other Squamish who held the licences could use one ten-fathom (eighteen-metre or sixty-foot) gill net as a set net – that is, in a fixed location; drift netting was prohibited. Beginning in June, they could fish for twenty-four hours each week, from 12 (noon) on Saturday until 12 (noon) on Sunday. From October through December, the fishing window opened for seventy-two hours a week from Thursday to Sunday. They were also required to mark their fishing gear and carry the licence while fishing, and there were restrictions on the material in the nets.

The Fisheries officer did not return the nets. Instead he charged Lewis with five counts under the *Fisheries Act* on grounds that Lewis had been using two nets where the licence permitted one; that one net was over the ten-fathom limit; that the nets were still in the water on Monday when the fishery was closed; that the nets were unmarked; and that the nets used prohibited material. Two other members of the Lewis family, Allen Frances and William Douglas, had been charged with similar offences the year before, and a fourth, Allen Jacob, was charged in 1987.

The four accused appeared before Judge Walker in the British Columbia Provincial Court in October 1987. The principal argument in *R. v. Lewis* concerned the boundary of the Cheakamus reserve. If the reserve extended into the river to the spot where the accused had been fishing, it would mean the fishing had occurred on the reserve. And if the fish had been caught on the reserve, then, under the terms of the *Indian Act*, a Squamish Indian Band by-law superseded the federal fisheries regulations. By-Law No.10, "for the preservation, protection and management of fish on the reserve," permitted members of the Squamish band to fish on Squamish reserves without the restrictions set out in the food fishing licence.[11] The location of the reserve boundary turned on the *ad medium filum aquae* presumption. Lawyers for the Squamish argued that the band owned the bed of the river, a non-tidal body of water, to its midpoint and therefore held the exclusive right to fish in its waters. However, Judge Walker ruled that, in Canada, the presumption did not apply in navigable waters; the Squamish River was navigable, and therefore the river was not a part of the reserve. The Lewises were bound by the terms of the Indian food fish licences, and Judge Walker levied fines of twenty-five dollars for each offence.

The Squamish appealed to the BC County Court where, in 1989, most of the convictions were overturned. Judge van der Hoop held that the alleged offences had occurred on the reserve. In doing so, he indicated that "the historical background of the right of the Indians to fish," and "the desire of both Provincial and Federal Governments to support and protect that right," were relevant in determining the boundaries of the reserve.[12] However, on the Crown's appeal of this decision, the BC Court of Appeal reinstated the convictions. The Squamish Indian Reserve No. 11 Cheakamus did not extend into the Squamish River, a finding that the Supreme Court of Canada upheld in 1996.[13]

In upholding the convictions, the Supreme Court relied on evidence that none of the lower courts had seen. It allowed an intervenor – Canadian National Railway – to submit new evidence in *Lewis* and in *R. v. Nikal*, a similar case involving Wet'suwet'en Indian reserve boundaries and fisheries on the Bulkley River, a tributary of the Skeena, that the court heard at the same time.[14] The new evidence consisted primarily of nineteenth-century correspondence to and from the Department of Marine and Fisheries regarding the rights of the public and of Native peoples to the fisheries. Largely on the basis of this evidence, which revealed the federal department's determined opposition to Native fishing rights, the Supreme Court concluded that the Indian reserve commissioners had not intended to allocate exclusive fisheries and, furthermore, that they did not have the authority to do so when they allotted Indian reserves. Justice Iacobucci, who wrote for the court in *Lewis,* concluded in regard to the Cheakamus reserve that "it was never the Crown's intention, at any point in time, to include a fishery as part of the reserve" and, more generally, "that the Crown's policy was to treat Indians and non-Indians equally as to the uses of the water and not to grant exclusive use of any public waters for the purpose of fishing."[15] In his summary he wrote: "It was never the intention of the Crown to provide the Bands with an exclusive fishery in waters adjacent to the reserves."[16] Similarly, Justice Cory, who wrote for the majority in *Nikal,* concluded that "the Crown in all its manifestations was consistently clear in its statements that no exclusive fishery should be granted to Indian bands in British Columbia."[17]

The Indian reserve commissions never formally linked Squamish Indian Reserve No. 11 Cheakamus to the fisheries. This much of the Supreme Court's decision in *Lewis* was correct. Covering more than 4,000 acres, the reserve was almost twice as large as all the other Squamish reserves combined (Figure 0.1). Forty years after the Joint Indian Reserve Commission allotted the Cheakamus reserve in 1876, the Royal Commission on Indian Affairs for the Province of British Columbia described small reserves above, immediately below, and across the Squamish River from the Cheakamus reserve as "fishing stations." The Squamish River was clearly an important fishing river for the Squamish people, but fishing was not the identifying feature of their largest reserve. Instead, the commissioners labelled the one Squamish reserve where it was possible to do something in addition to fishing as "pasturage and timber area."[18]

The Supreme Court's conclusions about the authority of the reserve commissioners and the intent of the Crown are more difficult to sustain. It is clear that the Department of Fisheries understood Native peoples had limited privileges to a food fishery. Beyond this privilege, Native fishers shared with the rest of the public the right not to be excluded from the fisheries, but nothing more. This view was well entrenched in the department when it assumed jurisdiction over the Pacific coast fisheries, and it is this understanding that pervades the decisions of the Supreme Court in *Nikal* and *Lewis.* It was, however, a relatively new understanding, at odds with the Crown's recognition of exclusive Native fisheries in the late eighteenth and early nineteenth centuries in Upper Canada. Indeed, the mid-nineteenth-century legal opinions on which Fisheries officials based their management of the Native fisheries, and on which the Supreme Court's recent judgments rest, amounted to a fundamental reinterpretation of fisheries law, policy, and treaty rights, the principal object of which was to diminish Native control over and access to the fisheries.[19]

Although Fisheries articulated its vision of limited Native fishing privileges most forcefully, it was never the only view, even in the late nineteenth century. Native peoples were clear that their rights to fish derived from their own legal traditions, not from the Crown, and those rights included, but were not limited to, fishing for food. The fisheries clause in the Douglas Treaties – the right to "fisheries as formerly" – provides some evidence that colonial officials in British Columbia also understood that Native fishing rights emanated from past practice, not the Crown, and that those rights were broadly construed.[20] Furthermore, there is clear and abundant evidence that, whether as a matter of law or policy, the Department of Indian Affairs sought to preserve Native peoples' access to their fisheries. Indeed, throughout the process of reserve allotments, Indian Affairs was a continuing, although faltering and occasionally ambivalent, advocate of Native fishing rights, at times insisting that Fisheries recognize the allocation of exclusive fisheries as part of the established reserve system or assume the costs of not doing so, while at other times deferring to its more powerful sibling department within the Dominion government. The prosecution of Squamish fisher Domanic Charlie, recounted at the outset of this book and in Chapter 6, reveals both the vigorous defence of Native fishing rights by Indian Affairs and the department's conflicted ambivalence towards those rights. It would oppose Fisheries, even in the public forum of the courtroom, but only so far as the county court, in *R. v. Charlie*. Nonetheless, Indian Affairs was a dissenting voice, albeit inconsistently expressed and seldom heard, within the government of Canada to Fisheries' refusal to recognize Native fishing rights. The Supreme Court's conclusions that "the Crown in all its manifestations was consistently clear that no exclusive fishery should be granted," or that "it was never the intention of the Crown to provide Bands with an exclusive fishery in waters adjacent to the reserves," are wrong. In concluding thus, the court has uncritically adopted the position of the Department of Marine and Fisheries and its interpretation of the law in the late nineteenth century as its own.

If, as I think must now be clear, the reserve geography in most of the province is premised on access to fish, then that geography ought to create certain entitlements to the fisheries. This is not an argument based on Aboriginal rights or title, although these underlie the claim, but instead on the government policy and practice of alloting small parcels of land connected to the fisheries. Indeed, the allotment of reserve land followed the identification of important fishing grounds, establishing a nexus between land and fish that should be recognized and restored.

This is not a novel idea. Peter Kelly, president of the Allied Tribes, put forward such a proposal in the 1920s. If Native peoples were to accept the 1916 report of the Royal Commission on Indian Affairs for the Province of British Columbia as the resolution of the Indian land question, Kelly insisted that a number of conditions be met, among them that Native peoples hold exclusive rights to the seine fisheries on reserve foreshores and in rivers running through reserves, and that certain other areas should be set aside for exclusive Native fisheries.[21] To support this proposition, he pointed to a decision of the United States Supreme Court that linked Indian reservations to the fisheries in Alaska.

In the late 1880s, a group of Metlakatla Tsimshian left northern British Columbia for Alaska, where the Annette Islands were set aside as their reservation. In 1916, a California company, Alaska Pacific Fisheries, erected a substantial fish trap extending out from one of the islands. The US government sued for its removal on the grounds that the reservation, which had been allotted to the Metlakatla on the understanding that they would move their salmon canning operation north from British Columbia to Alaska, included not only the islands but the adjacent fisheries as well. In *Alaska Pacific Fisheries v. United States,* the US Supreme Court found that the trap, which would catch an estimated 600,000 salmon, would "tend materially to reduce the natural supply of fish accessible to the Indians," and the court ordered its removal.[22] In the United States, where an Indian reservation had been located to secure access to the fisheries, the reservation included the fisheries: "The Indians could not sustain themselves from the use of the upland alone. The use of the adjacent fishing grounds was equally essential. Without this the colony could not prosper in that location. The Indians naturally looked on the fishing grounds as part of the islands and proceeded on that theory in soliciting the reservation."[23]

These observations could apply equally to a great many reserves in British Columbia. Indeed, in a case involving the prosecution of Cowichan fishers in the 1970s, Supreme Court of Canada Justice Brian Dickson concluded as much. "It is extremely difficult," he wrote in *R. v. Jack,* "to separate the fishery from either Indians or the lands to be reserved for Indians. In the latter case, lands were to be reserved to Indians for the purpose of permitting them to continue their river fishery at the customary stations. In the former case, the Indians were to be encouraged to exploit the fishery, for both their own benefit and that of the incoming white settlers, as a means of avoiding the Indians becoming a charge upon the colonial finances."[24] The Supreme Court omitted a reference to this decision in *Nikal* and *Lewis*. It considered *Alaska Pacific Fisheries* briefly in *Lewis*, but distinguished it on the grounds that it involved "islands and intervening waters," not a river.[25] The court chose to focus on the Department of Fisheries' insistence that there would be no exclusive Native fisheries, rather than on the purposes for which reserves were allotted.[26]

It is impossible to ignore the continuing conflict over fish in British Columbia, which looms over the writing of a history such as this. Although detached from the litigation, I have written this book into that conflict. Unease lingers over the fisheries in this province, fuelled by concerns about their long-term sustainability and the lack of resolution of Aboriginal rights and title. This unease is reflected in the conflict that has simmered through my writing, and it continues to haunt a society that is still coming to terms with its colonial past. The different conceptions of rights to fish have not led to the eruption that occurred on the Atlantic coast in 1999 following the Supreme Court of Canada decision in *R. v. Marshall*, which confirmed Mi'kmaq treaty rights to commercial fisheries.[27] But every July and August, as sockeye migrate to spawning grounds, the possibility that the persistent unrest in the fishery on the Pacific will become more volatile seems to grow. Conflict has enveloped other fisheries as well, but the salmon fleet is the largest on the coast, and the diversity of claims, gear types, and locations, as well as the challenge posed by the growing aquaculture industry and the apparent stresses of a changing climate, make it the most difficult to manage and the most contentious. If fish "disappear," as they seem to have done most recently in 2004, when sockeye that were counted in the lower Fraser did not arrive on the spawning grounds in expected numbers, fingers are quickly pointed and Fisheries is blamed for not enforcing the law against Native fishers. And whether fish "disappear" or not, Native fishers continue to appear regularly in provincial courthouses to respond to alleged *Fisheries Act* violations.

While the constitutional priority of an Aboriginal food fishery has been generally accepted, at least in principle if not always in practice, Aboriginal or treaty rights to commercial fisheries remain strongly contested. In the last few years, members of the commercial fishing fleet have launched protest fisheries in an effort to put Department of Fisheries and Oceans' management strategies on trial, particularly the "pilot sales program" that provided limited priority for a few First Nations to catch fish for sale. Their argument has been that these Aboriginal fisheries violate the public right to fish[28] and the equality guarantees in the constitution.[29] The sales program was intended as a temporary measure while treaties were negotiated.

Treaty agreements have proven elusive, however, in part because of difficult negotiations over fish. Governments have been reluctant to include fisheries allocations in the treaties themselves, insisting that any such provisions be included in side agreements known as "harvest agreements."[30] These arrangements reflect the anxiety in the commercial fleet, which itself holds uncertain tenure to the fisheries, about the constitutional entrenchment of Aboriginal fishing rights. They also indicate the continuing salience of the common-law doctrine of the public right to fish and the residual reluctance to recognize anything that might be construed as an exclusive First Nations' fishery.

Although treaty negotiations have been slow to produce treaties, and court cases have been inconclusive, it is clear that these processes will provide Native peoples in British Columbia with greater access to and control over the province's land and resources than they have had since the imposition of Indian reserves and Canadian fisheries law in the late nineteenth century. The principal legal basis for this transformation lies in the growing recognition of Aboriginal rights and title as legal interests that confer rights to land and resources, including fisheries. The Indian reserve geography of British Columbia, premised on access to the fisheries, may also be understood as evidence of those rights; the refusal to recognize the connections, indicative of their denial. That geography should also be seen as one of the grounds that support a renewed Native presence in the fisheries.

Appendix

Indian Reserves Allotted for Fishing Purposes in British Columbia, 1849-1925

The maps in the Appendix indicate the names and locations of the nearly 750 reserves in British Columbia that government officials (primarily Indian reserve commissioners) allotted before 1925 for fishing purposes. These purposes ranged from the common designations of a reserve as a "fishing station," "fishing place," or simply "fishery," to a "camping place" or "well-sheltered nook" used on route to fishing grounds, or to "drying grounds" and particular locations that contributed wood or other materials used in the processing of fish. Many of the reserves were noted to be "worthless," "unsuitable," or "of little value" for any purpose other than fishing. Some reserve grants included specific reference to the "right to fish" or the "exclusive right to fish" in waters adjacent to or in proximity to reserved land. The reserved fisheries included salmon, halibut, dogfish, herring, oolichan, bass, trout, whitefish, lamprey eel, shark, and "deep sea fisheries." They also included reserves set aside for the hunting of seals and otters, for the gathering of clams and cockles, and the harvesting of seaweed.

Some of the details of these Indian reserves, including the allotment dates, the person or body responsible for the allotment, and the recognized fishing purpose are included in an accompanying table available in UBC's Information Repository, at http://hdl.handle.net/2429/648. The table includes additional reserves that were set aside by an Indian reserve commissioner but never confirmed by the provincial and federal governments. The Appendix maps include only those reserves that were confirmed, although some might have been subsequently cut-off or surrendered.

The maps and the table do not include the reserves in the northeast corner of the province allotted under Treaty 8.

Eric Leinberger, cartographer in the Department of Geography, UBC, has drawn the maps in the Appendix and throughout the book.

A1

Alexis Creek
Puntzi Lake 2

Xeni Gwet'in
Bunmia Lake 5
Lohbiee 3
Chilco Lake 1

Klahoose 1
Quaniwsom 2
Klahoose

Homalco 2
Homalco
Homalco 1
Orford Bay 4
Bear Bay 8
Salmon Bay 3
Deep
Valley 5

Bella Coola 1

Noosseeck 2

Nuxalk

Taleomy 3

Tsaawati 1
Da'naxda'xw
Sim Creek 5
Ahnuhati 6
We Wai
Kum
Kwiakah
Sahyouck 6
Keogh 2
Kwatse 3
Homayno 2
Matsayno 5
Pawala 5
Neville 4
Matlaten 4
Port
Loughborough 3

Echo 6
Clatse 5
Haanees 2
Kokyet 1
Dzle Island 14
Island 14A
Kunsoot 9
Kajustis 10
Howeet 8
Kisameet 7

Yellertlee 12
Werkinellelk 11

Kiltala 2
Katit 1
Oweekeno
Nekite 2
Tseetsum-Sawlatalah 6
Halowis 1
Gwa'sala-Nakwaxda'xw
Tsal-kwi-ee 13
Kor-kwi-ss 14
Waump 16
Oom-wo-too-a-wan 10
Pelhooth-tkai 17
Kwetlahks
Dove Island 12
Pahas
Matpahkum 4
Saagoombahlah'o
Ta-a-ack 5

Cockmi 3
Ann Island 7

Wawatti 12
Paleece 1
Kuthlo 18
Nalco 8
Quay 4
Gieyka 5
Lawanth 5
Kunstamis 2
Kadis 1

Tsawatineuk
Quee 7
Ahta 3
Kalkweken 4

Kwicksutaineuk-ah
-kwaw-ah-mish
Kwakwaxtala 10
Hopetown 10A
Meetup 2

Dakiulls 7
Kwiyapa 5
Apsagayu 1A
Umdagitis 9
Compton
Kwe' Qwa'Sot'Em
Mamalilikulla-
Karlukwees 1
Karlukwees 4A
Aglagumna 4
Tlowitsis-
Mumtagila
Haynett 6
Hayalite 3

Hope Island 1
Nahwitti 4
Wekems 6
Glen-gla-ouch 5

Ouchton 3
Semach 2
Tlatlasikwala
Tsulquate 4
Pa-Cart-lin-ne 3
Keogh 4
Ah-wei-chia-ol-to 16
Toh-quo-eugh 2
Cluena 14A
Koprino 12
Quatieyo 12
O-ya-kum-la 11
Mah-te-nicht 8
Quatsino

Gwawaenuk
Ches-la-kee 4
O-tsaw-las 5
Klicksewy
Klicdekdurna 7
Quatum 9
Lioltah 4
Teeta 7
Kwakiutl
Keogh 6
Cayuse 6
Namgis
Ar-ce-xiy-ee 4
Island 6
Doarf Point 5
Ku-la-das 6
Deer Island 1
Excelsti 1

Salmon
River 1

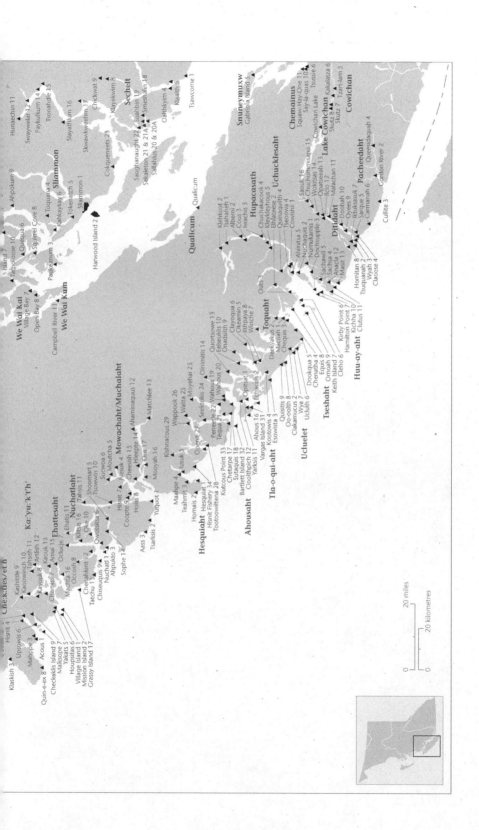

Stone
Stone 1

Saddle Horse 2

Soda Creek
Soda Creek 1

Tillion 4

Williams Lake
San Jose 6
Chimney Creek 5

Toosey
Toosey 3

Esketemc
Windy Mouth 7

Dog Creek 4

Canoe Creek
Fish Lake 5

High Bar
High Bar 1

Ts'kw'aylaxw
Leon Creek 2

Whispering
Pines
Kelly Creek 3

Nekaliston 2

Bridge River
Bridge River 1
Bridge River 2

Fountain
Dry Salmon 7

Cayoose Creek
Cayoosh Creek 1

Seton Lake
Slilcon 2

Seton Lake 3
Seton Lake 4
Seton Lake 5

Mission 5
Necait 6

T'it'q'et
Pashilqua 2

N'Quatqua
Nequatque 1

Mount
Currie
Lokla 4

Barriere River 3
Louis Creek 5

North Thompson

Adams Lake

Squaam 2

Toopj 3

Spallumcheen
Sicamous 3

Switsemalph 7

Okanagan
Swan Lake 4

Priest's Valley 6

Kamloops 5

Salmon Lake 7

Chaperon

Kamloops

Kamloops 2
Kamloops 3

Hihium Lake 6

Bonaparte

Leon Lake 4

Mauvais Rocher 5

Pipseul 5

105 Mile Post 2
Cheetsum's Farm 1

Ashcroft
Oregon Jack Creek 5

Oregon Jack Creek

Upper Tsinkahtl 8A
Tsinkahtl 8

Shackan
Shackan 11

McCartney's Flat 4
Fish Lake 7

Cook's
Ferry
Skoonkoon 2

Nicomen

Lytton
Neskkep 6
Seah 5

Fish Lake 7

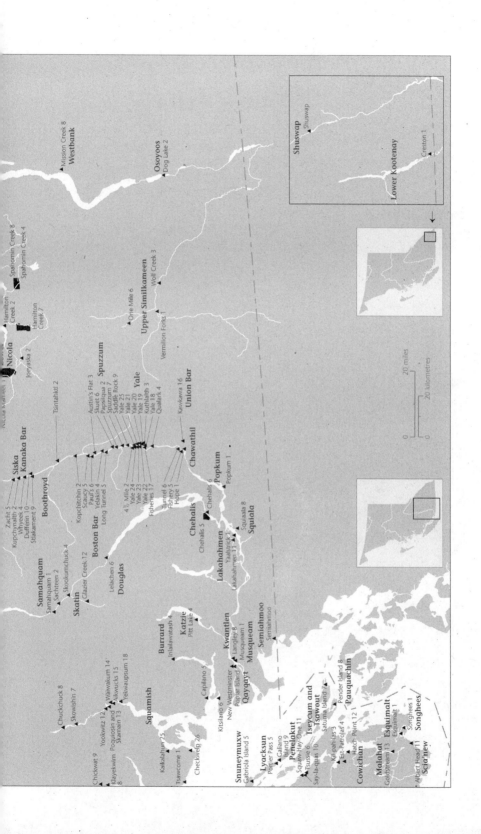

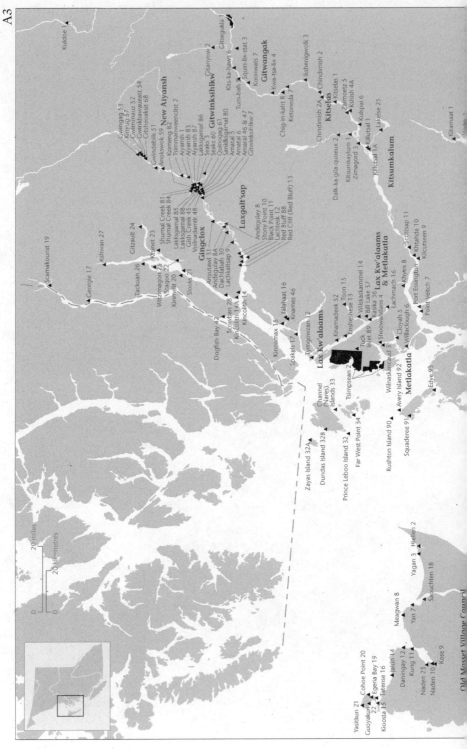

A3

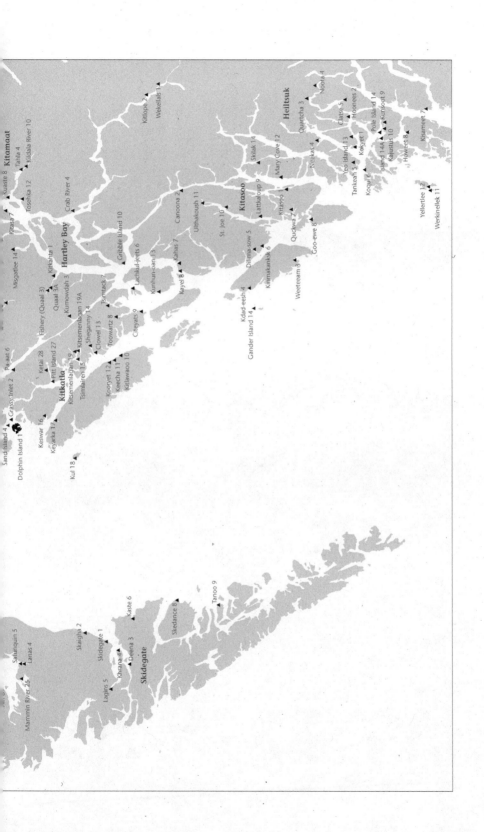

A4

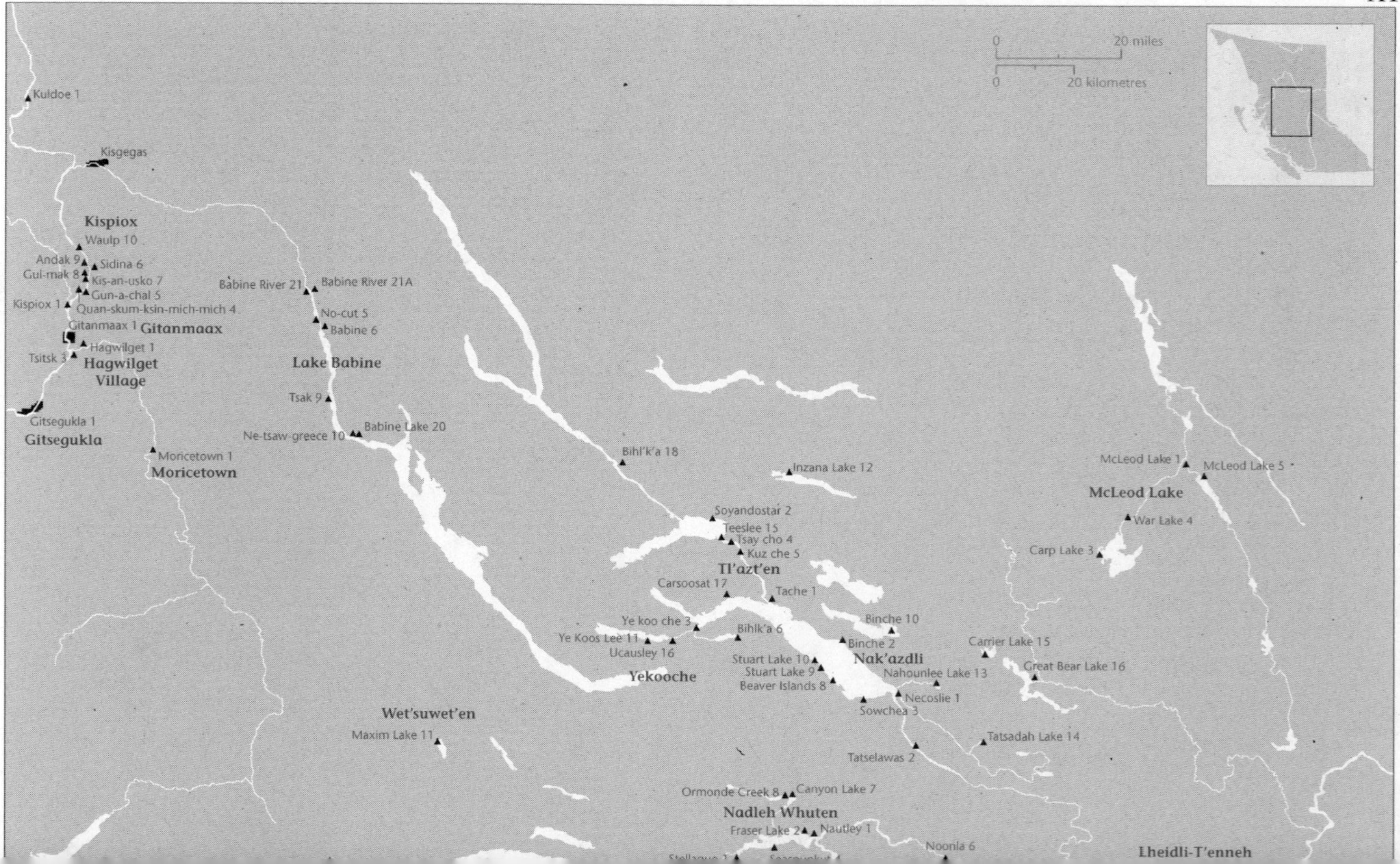
0
20 miles
0
20 kilometres
Kuldoe 1
Kisgegas
Kispiox
Waulp 10
Andak 9
Sidina 6
Gul-mak 8
Kis-an-usko 7
Gun-a-chal 5
Kispiox 1
Quan-skum-ksin-mich-mich 4
Gitanmaax 1
Gitanmaax
Hagwilget 1
Tsitsk 3
Hagwilget Village
Gitsegukla 1
Gitsegukla
Moricetown 1
Moricetown
Babine River 21
Babine River 21A
No-cut 5
Babine 6
Lake Babine
Tsak 9
Ne-tsaw-greece 10
Babine Lake 20
Bihl'k'a 18
Inzana Lake 12
Soyandostar 2
Teeslee 15
Tsay cho 4
Kuz che 5
Tl'azt'en
Carsoosat 17
Tache 1
Ye koo che 3
Ye Koos Lee 11
Ucausley 16
Bihlk'a 6
Yekooche
Binche 10
Binche 2
Stuart Lake 10
Stuart Lake 9
Beaver Islands 8
Nak'azdli
Nahounlee Lake 13
Necoslie 1
Sowchea 3
Carrier Lake 15
Great Bear Lake 16
Tatsadah Lake 14
Tatselawas 2
Wet'suwet'en
Maxim Lake 11
Ormonde Creek 8
Canyon Lake 7
Nadleh Whuten
Fraser Lake 2
Nautley 1
Noonla 6
McLeod Lake 1
McLeod Lake 5
McLeod Lake
War Lake 4
Carp Lake 3
Lheidli-T'enneh

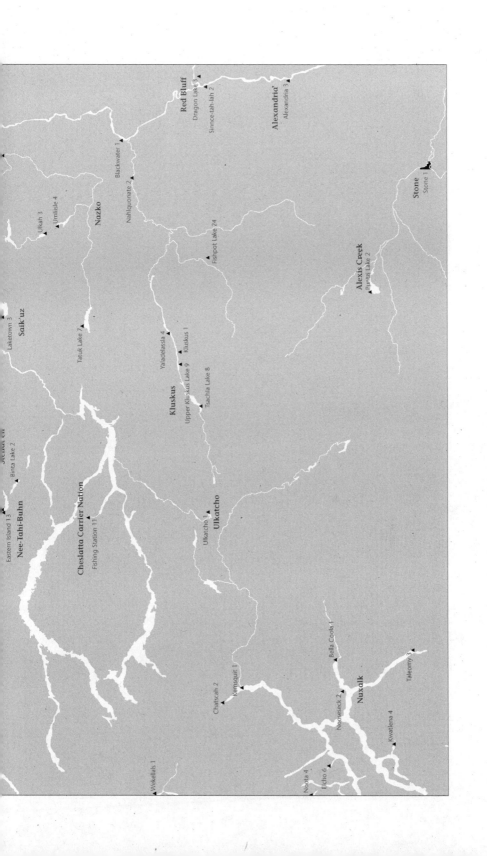

Red Bluff
Dragon Lake 3
Sinnce-tah-lah 2

Alexandria'
Alexandria 3

Stone
Stone 1

Blackwater 1

Nahlquonate 2

Nazko

Ulkah 3
Umtiside 4

Soik'uz

Laketown 3

Tatuk Lake 7

Fishpot Lake 24

Alexis Creek
Puntzi Lake 2

Kluskus
Yaladelasla 4
Upper Kluskus Lake 9
Kluskus 1
Tsacha Lake 8

Binta Lake 2

Nee-Tahi-Buhn
Eastern Island 1 3

Cheslatta Carrier Nation
Fishing Station 11

Ulkatcho 3
Ulkatcho

Chatscah 2

Kensquit 1

Bella Coola 1

Wakedlals 1

Nuxalk
Noosseick 2
Kwatlena 4
Taleomy 3

Noota 4
Elcho 6

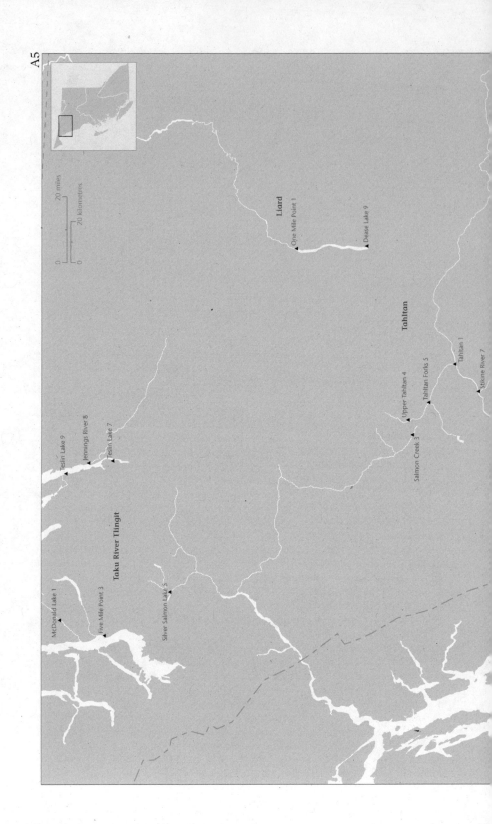

Notes

Acknowledgments

1 The five cases from British Columbia were *R. v. Van der-Peet*, [1996] 2 S.C.R. 507, [1996] 4 C.N.L.R. 177; *R. v. Gladstone*, [1996] 2 S.C.R. 723, [1996] 4 C.N.L.R. 65; *R. v. N.T.C. Smokehouse Ltd.*, [1996] 2 S.C.R. 672, [1996] 4 C.N.L.R. 130; *R. v. Nikal*, [1996] 1 S.C.R. 1013, [1996] 3 C.N.L.R. 178; *R. v. Lewis*, [1996] 1 S.C.R. 921, [1996] 3 C.N.L.R. 131. The two from Quebec were *R. v. Adams*, [1996] 3 S.C.R. 101, [1996] 4 C.N.L.R. 1, and *R. v. Côte*, [1996] 3 S.C.R. 139, [1996] 4 C.N.L.R. 26.
2 *R. v. Sparrow*, [1990] 1 S.C.R. 1075, [1990] 3 C.N.L.R. 160.

Introduction

1 For a brief history of Domanic Charlie's life see Oliver N. Wells, ed., *Squamish Legends by Chief August Jack Khahtsahlano and Domanic Charlie* (Vancouver: Charles Chamberlain and Frank T. Coan, 1966). On the mixed industrial/traditional economies see John Lutz, "Seasonal Rounds in an Industrial World," in *A Stó:lō -Coast Salish Historical Atlas*, ed. Keith Thor Carlson (Vancouver: Douglas and McIntyre, 2001), 64-67; Robert Galois, "Colonial Encounters: The Worlds of Arthur Wellington Clah," *BC Studies* 115-16 (1997-98): 105-47. On the changing Squamish economy see Chris Roine, "The Squamish Aboriginal Economy, 1860-1940" (master's thesis, Department of History, Simon Fraser University, 1996).
2 On Squamish use of marine plants and animals, including a description of gaff hooks, see Dorothy I.D. Kennedy and Randy Bouchard, *Utilization of Fish, Beach Foods, and Marine Mammals by the Squamish Indian People of British Columbia* (Victoria: British Columbia Indian Language Project, 1976), archived at UBC Library, Special Collections. On the Coast Salish see Wayne Suttles, *Coast Salish Essays* (Vancouver: Talonbooks, 1987); Bruce G. Miller, ed., *Be of Good Mind: Essays on the Coast Salish* (Vancouver: UBC Press, 2007).
3 I generally use the terms Native peoples, Native title, Native rights rather than Aboriginal. The latter, although more common now, was not much used in the period studied. I use Aboriginal rights, title, etc., when referring to recent legal developments. I use the terms Indian band, Indian reserve, Indian food fishery, etc., when referring to the particular legal category or status accorded by the state.
4 Thomas Wilson Lambert, *Fishing in British Columbia* (London: H. Cox, 1907), 70. For a history of the river, including Native and sport fisheries, see James W. Morton, *Capilano: The Story of a River* (Toronto: McClelland and Stewart, 1970).
5 See the cases of *R. v. Nikal*, [1996] 1 S.C.R. 1013, [1996] 3 C.N.L.R. 178, and *R. v. Lewis*, [1996] 1 S.C.R. 921, [1996] 3 C.N.L.R. 131, discussed in the conclusion. See

also the very different result from the United States Supreme Court in *Alaska Pacific Fisheries v. United States* (1918), 248 U.S. 78. The Supreme Court of Canada has mentioned the link between Indian reserves and the fisheries in British Columbia in a number of other cases, including *R. v. Jack*, (1979) [1980] 1 S.C.R. 294 at 308, [1979] 2 C.N.L.R. 25; and *Wewaykum Indian Band v. Canada*, [2002] 4 S.C.R. 245; [2003] 1 C.N.L.R. 341, at para. 11.

6   Cole Harris, *Making Native Space: Colonialism, Resistance, and Reserves in British Columbia* (Vancouver: UBC Press, 2002), notes the importance of fisheries in the process of allotting reserve land but does not provide sustained analysis. Earlier and important published work on colonial land policy that ignores the fisheries includes: G.E. Shankel, "The Development of Indian Policy in British Columbia" (PhD diss., University of Washington, 1945); Wilson Duff, *The Impact of the White Man*, 3rd ed. (Victoria: Royal British Columbia Museum, 1997); Robert E. Cail, *Land, Man, and the Law: The Disposal of Crown Lands in British Columbia, 1871-1913* (Vancouver: UBC Press, 1974); Robin Fisher, *Contact and Conflict: Indian-European Relations in British Columbia, 1774-1890*, 2nd ed. (Vancouver: UBC Press, 1992); Paul Tennant, *Aboriginal People and Politics: The Indian Land Question in British Columbia, 1849-1989* (Vancouver: UBC Press, 1990).

7   Here I would include my own work, Douglas C. Harris, *Fish, Law, and Colonialism: The Legal Capture of Salmon in British Columbia* (Toronto: University of Toronto Press, 2001). Dianne Newell, *Tangled Webs of History: Indians and the Law in Canada's Pacific Coast Fisheries* (Toronto: University of Toronto Press, 1993), 55-62, 87-88, notes the connection between land policy and the fisheries, and points out that historians have paid little attention to how access to and disputes over fish affected the allotment of reserves.

8   C. Harris, *Making Native Space*.

9   On the history of Native title in British Columbia see Hamar Foster, "Letting Go the Bone: The Idea of Indian Title in British Columbia, 1849-1927," in *Essays in the History of Canadian Law*, vol. 6, *British Columbia and the Yukon*, ed. Hamar Foster and John McLaren (Toronto: Osgoode Society for Canadian Legal History, 1995), 28-86.

10   *Delgamuukw v. British Columbia*, [1997] 3 S.C.R. 1010, [1998] 1 C.N.L.R. 14; *Tsilhqot'in Nation v. British Columbia*, 2007 B.C.S.C. 1700.

11   In *Haida Nation v. British Columbia (Minister of Forests)*, 2004 SCC 73, [2004] 3 S.C.R. 511, and *Taku River Tlingit v. British Columbia (Project Assessment Director)*, 2004 SCC 74, [2004] 3 S.C.R. 550, the court set out the duty of the Crown to consult and accommodate Aboriginal interests when it contemplates activity that has the potential to infringe claimed, but not yet proven, Aboriginal rights of title.

12   C. Harris, *Making Native Space*, 261. See the maps of the reserve geographies in neighbouring jurisdictions in the introduction, xxv-xxvi.

13   Field Minutes, 20 October 1886, in Federal Collection of Minutes of Decision, Correspondence and Sketches, vol. 11 (P. O'Reilly [Indian Reserve Commissioner], June 1885-March 1889, file no. 29858, vol. 5, file 20802 [extracts], [Reg. No. B-64646]), p. 248 (copy held by Department of Indian Affairs and Northern Development, Vancouver Regional Office).

14   Some of the important book-length works include Hilary Stewart, *Indian Fishing: Early Methods on the Northwest Coast* (Vancouver: J.J. Douglas, 1977); Reuben M. Ware, *Five Issues, Five Battlegrounds: An Introduction to the History of Indian Fishing in British Columbia* (Chilliwack, BC: Coqualeetza Education Training Centre, 1983); Newell, *Tangled Webs*; Daniel Boxberger, *To Fish in Common: The Ethnohistory of Lummi Indian Salmon Fishing* (Lincoln: University of Nebraska Press, 1989; Seattle: University of Washington Press, 2000); Joseph E. Taylor III, *Making Salmon: An Environmental History of the Northwest Fisheries Crisis* (Seattle: University of Washington Press, 1999).

15 See J. Michael Thoms, "Ojibwa Fishing Grounds: A History of Ontario Fisheries Law, Science, and the Sportsmen's Challenge to Aboriginal Treaty Rights, 1650-1900" (PhD diss., Department of History, University of British Columbia, 2004), 76-78. P.B. Munsche, *Gentlemen and Poachers: The English Game Laws 1671-1831* (Cambridge: Cambridge University Press, 1981), makes a similar argument in respect of game.

16 A.C. Anderson, "Notes on the Indian Tribes of British North America, and the Northwest Coast," *Historical Magazine* 7 (March 1863): 80, http://www.canadiana.org/ECO/ItemRecord/16598?id=3975306651140155.

17 This includes reserves allotted or confirmed by the Joint Indian Reserve Commission (1876-78), the Indian Reserve Commission (1880-1910), and the Royal Commission on Indian Affairs for the Province of British Columbia (1913-16) as amended by the Ditchburn Clark Agreement (1924). It does not include the reserves in northeastern British Columbia allotted under Treaty 8.

18 A complete list of these reserves, organized by First Nation and with information about the person or body who allotted the reserve, the date of allotment or confirmation, and the document identifying the connection to the fisheries, is available online in UBC's Information Repository, at http://hdl.handle.net/2429/648.

19 Even theorists such as Edward Said and Robert J.C. Young, whose work has focused on the cultural and discursive power of imperialism, have recognized the material conflict over land at its root. According to Said, "Imperialism means thinking about, settling on, controlling land that you do not possess, that is distant, that is lived on and owned by others." *Culture and Imperialism* (New York: Vintage Books, 1994), 7. Young wrote that "colonialism above all involves the physical appropriation of land." *Colonial Desire: Hybridity in Theory, Culture and Race* (London: Routledge, 1995), 172.

20 For surveys of this literature see Sally Engle Merry, "Law and Colonialism," *Law and Society Review* 25 (1991): 889-922; Jeffrey R. Dudas, "Law at the American Frontier," *Law and Social Inquiry* 29 (2004): 859-90.

21 See John Gerard Ruggie, "Territoriality and Beyond: Problematizing Modernity in International Relations," *International Organization* 47, 1 (1993): 139-74.

22 For an analysis of the changing notion of sovereignty, particularly in relation to indigenous peoples, see P.G. McHugh, *Aboriginal Societies and the Common Law: A History of Sovereignty, Status, and Self-Determination* (Oxford: Oxford University Press, 2004).

23 James C. Scott, *Seeing like a State: How Certain Schemes to Improve the Human Condition Have Failed* (New Haven, CT: Yale University Press, 1998), 36.

24 E.P. Thompson, *Whigs and Hunters: The Origin of the Black Act* (New York: Pantheon Books, 1975), and *Customs in Common: Studies in Traditional Popular Culture* (New York: The New Press, 1993), particularly "Custom, Law, and Common Right." See also Douglas Hay and Nicholas Rogers, *Eighteenth-Century English Society: Shuttles and Swords* (Oxford: Oxford University Press, 1997), chapters 6 and 7.

25 See Theodore F.T. Plucknett, *A Concise History of the Common Law,* 5th ed. (Boston: Little, Brown, 1956), 277-89. On Coke's influence see Nicholas K. Blomley, *Law, Space, and the Geographies of Power* (New York: The Guilford Press, 1994), 76-105.

26 The centrality of the right to exclude to the idea of property is much debated. For a sampling of that literature see Felix Cohen, "Dialogue on Private Property," *Rutgers Law Review* 9 (1954): 357; T.C. Grey, "The Disintegration of Property," in *Property: Nomos 12,* ed. J.R. Penncock and J.W. Chapman (New York: New York University Press, 1980), 69; Thomas W. Merrill, "Property and the Right to Exclude," *Nebraska Law Journal* 77 (1998): 730-55.

27 Jeremy Waldron, *The Right to Private Property* (Oxford: Clarendon Press, 1988), 42.

28 William Blackstone, *Commentaries on the Laws of England,* a facsimile of the first edition of 1765-69 (Chicago: University of Chicago Press, 1979), 2:2.

29 Joseph Singer, *Entitlement: The Paradoxes of Property* (New Haven, CT: Yale University Press, 2000), 3.
30 Political scientist C.B. MacPherson has argued that, "whereas in pre-capitalist society property was understood to comprise common as well as private property, with the rise of capitalism the idea of common property drops virtually out of sight and property is equated with private property – the right of a natural or artificial person to exclude others from some use or benefit of something." "Capitalism and the Changing Conception of Property," in *Feudalism, Capitalism and Beyond*, ed. Eugene Kamenka and R.S. Neale (Canberra: ANU Press, 1975), 105. See also C.B. MacPherson, *Property: Mainstream and Critical Positions* (Toronto: University of Toronto Press, 1978), 1-13.
31 Imperial powers also deployed regimes of private property in administrative colonies to govern territory and control people. Ranajit Guha, *A Rule of Property for Bengal: An Essay on the Idea of Permanent Settlement* (Paris: Mouton, 1963; Durham, NC: Duke University Press, 1996), describes the British attempt in India to confirm the position of a landowning class as an integral part of colonial rule.
32 For an account of the different methods used by European powers in an effort to validate or legitimize their claims to possession see Patricia Seed, *Ceremonies of Possession in Europe's Conquest of the New World, 1492-1640* (Cambridge: Cambridge University Press, 1995).
33 "Treaty Establishing the Boundary in the Territory on the Northwest Coast of America Lying Westward of the Rocky Mountains (Oregon Treaty)," United States and United Kingdom, 15 June 1846, 9 U.S. Stat. 869, F.O 94/372, P. 722 (1846) LII 185.
34 Daniel Clayton, *Islands of Truth: The Imperial Fashioning of Vancouver Island* (Vancouver: UBC Press, 2000), 234-35.
35 Scott, *Seeing like a State*, 36.
36 Robert David Sack, *Human Territoriality: Its Theory and History* (Cambridge: Cambridge University Press, 1986), 33-34, 63, emphasizes the ways in which territorial strategies construct spaces that are "emptiable and fillable."
37 This was one of the principal functions of the *Royal Proclamation of 1763*, R.S.C. 1985, App. 2, No. 1. As an interest in "fee," it was an inheritable interest that was "simple" in the sense that its holder could devise it to anyone. All land in the common-law system was held of the Crown; "free and common socage" described the nature of that tenurial relationship. Originally a form of agricultural tenure that required its holder to provide the Crown with agricultural products or labour, or money in lieu of such services, by the nineteenth century these incidents of tenure had long been removed, leaving only the foundational tenurial relationship without the associated obligations. Apart from the land being held of the Crown, the notion of tenure had been stripped of its feudal trappings. Extricated from a web of personal relationships, it had been radically simplified and, as such, was centrally known or knowable.
38 Cole Harris, "How Did Colonialism Dispossess? Comments from an Edge of Empire," *Annals of the Association of American Geographers* 94 (2004): 165, describes law, along with maps and numbers, as the principal instruments of the colonial state that effect dispossession.
39 John Comaroff, "Colonialism, Culture, and the Law: An Introduction," *Law and Social Inquiry* 26 (2001): 309.
40 Paige Raibmon, "Un Making Native Space: A Genealogy of Indian Policy, Settler Practice and the Micro-Techniques of Dispossession," in *The Power of Promises: Perspectives on Treaties with Native Peoples in the Pacific Northwest*, ed. Alexandra Harmon (Seattle: University of Washington Press, forthcoming 2008).
41 On the relationship between sovereignty and property see Morris Cohen, "Property and Sovereignty," *Cornell Law Quarterly* 13 (1927-28): 8-30; and Joseph Singer, "Sovereignty and Property," *Northwestern University Law Review* 86, 1 (1991-92): 1-56.

42 Jeremy Bentham, "Community of Goods – Its Inconveniences," Part 2, Chapter 6, of "Principles of the Civil Code," cited in A.V. Dicey, "The Paradox of the Land Law," *Law Quarterly Review* 21 (1905): 226.

43 *Indian Act*, S.C. 1876, c. 18.

44 Douglas Sanders, "Private Property on Indian Reserves" (unpublished manuscript in possession of the author, 14 January 2001); Christopher Alcantara, "Individual Property Rights on Canadian Indian Reserves: The Historical Emergence and Jurisprudence of Certificates of Possession," *Canadian Journal of Native Studies* 23 (2003): 391-424.

45 In 1866, the colony passed legislation requiring that "Aborigines of this Colony or the Territories neighbouring thereto" secure special permission from the governor in writing before they could pre-empt land; *Land Ordinance* (1866), R.S.B.C., 1871, c. 24. In 1907, the province passed legislation requiring that "aborigines of the continent" secure a special order from the lieutenant governor before they could purchase Crown land; *An Act to Amend the Land Act*, S.B.C. 1907, c. 25, s. 9(13).

46 Cornelius van Bynkershoek, *De Dominio Maris Dissertatio – The Dissertation on the Sovereignty of the Sea*, trans. Ralph Magoffin, ed. James Brown Scott (1744; New York: Oxford University Press, 1916), 43-44. See also Lea Brilmayer and Natalie Klein, "Land and Sea: Two Sovereignty Regimes in Search of a Common Denominator," *New York University Journal of International Law* 33 (2000-1): 703-68.

47 Hugo Grotius, *The Freedom of the Seas, or, the Right Which Belongs to the Dutch to Take Part in the East Indian Trade*, trans. Ralph Magoffin, ed. James Brown Scott (1608; New York: Oxford University Press, 1916), 30.

48 Ibid., 29.

49 I use the term "common property" in the limited sense of conferring a right not to be excluded. I am not using it to describe the rights in a commons, where members of an identified community hold defined and overlapping rights of use. Instead, I am using the term to describe the rights in what amounts to an open-access regime.

50 See Chapter 4 of this book.

51 See the analysis in J. Michael Thoms, "A Place Called Pennask: Fly-Fishing and Colonialism at a British Columbia Lake," *BC Studies* 133 (2002): 69-98.

52 Sproat's attempts were futile because the province refused to recognize his many detailed compromises, but also because these compromises could not address the unresolved issue of Native title.

53 Alicja Muszynski, *Cheap Wage Labour: Race and Gender in the Fisheries of British Columbia* (Montreal and Kingston: McGill-Queen's University Press, 1996).

54 Keith Thor Carlson, "Innovation, Tradition, Colonialism, and Aboriginal Fishing Conflicts in the Lower Fraser Canyon," in *New Histories for Old: Changing Perspectives on Canada's Native Pasts*, ed. Ted Binnema and Susan Neylan (Vancouver: UBC Press, 2007), 145-74.

55 See Kenneth Brealey, "First (National) Space: (Ab)original (Re)mappings of British Columbia" (PhD diss., Department of Geography, University of British Columbia, 2002).

Chapter 1: Treaties, Reserves, and Fisheries Law

1 On the early trade see Richard Mackie, *Trading beyond the Mountains: The British Fur Trade on the Pacific, 1763-1843* (Vancouver: UBC Press, 1996); Cole Harris, *The Resettlement of British Columbia: Essays on Colonialism and Geographical Change* (Vancouver: UBC Press, 1997), 34-42. On the set of relationships between imperial centre and periphery in the late nineteenth century see Daniel Clayton, *Islands of Truth: The Imperial Fashioning of Vancouver Island* (Vancouver: UBC Press, 2000).

2 "Treaty Establishing the Boundary in the Territory on the Northwest Coast of America Lying Westward of the Rocky Mountains (Oregon Treaty)," United States and United Kingdom, 15 June 1846, 9 U.S. Stat. 869, F.O. 94/372, P. 722 (1846) LII 185.

3 See Jean Barman, *The West beyond the West: A History of British Columbia*, rev. ed. (Toronto: University of Toronto Press, 1996); C. Harris, *Resettlement of British Columbia*. On the early settlement of Vancouver Island see Richard Mackie, "The Colonization of Vancouver Island, 1849-1858," *BC Studies* 96 (1992-93): 3-40.

4 This point is made most forcefully by Robin Fisher, *Contact and Conflict: Indian-European Relations in British Columbia, 1774-1890*, 2nd ed. (Vancouver: UBC Press, 1992).

5 Daniel P. Marshall, "Claiming the Land: Indians, Goldseekers, and the Rush to British Columbia" (PhD diss., Department of History, University of British Columbia, 2000).

6 On the distribution of Crown lands in British Columbia see Robert E. Cail, *Land, Man, and the Law: The Disposal of Crown Lands in British Columbia, 1871-1913* (Vancouver: UBC Press, 1974). For a local study see R.W. Sandwell, *Contesting Rural Space: Land Policy and Practices of Resettlement on Saltspring Island, 1859-1891* (Montreal and Kingston: McGill-Queen's University Press, 2005). On the impact of fences see William Cronin, *Changes in the Land: Indians, Colonists, and the Ecology of New England* (New York: Hill and Wang, 1983). Also George Manuel and Michael Posluns, *The Fourth World: An Indian Reality* (Don Mills, ON: Collier-Macmillan, 1974), 48-53.

7 Nicholas K. Blomley, "Law, Property, and the Geography of Violence: The Frontier, the Survey, and the Grid," *Annals of the Association of American Geographers* 93 (2003): 124.

8 John Weaver, *The Great Land Rush and the Making of the Modern World, 1650-1900* (Montreal and Kingston: McGill-Queen's University Press, 2003), 18.

9 Patricia Seed, *Ceremonies of Possession in Europe's Conquest of the New World, 1492-1640* (Cambridge: Cambridge University Press, 1995).

10 Blomley, "Law, Property," 128-29.

11 Cole Harris, *Making Native Space: Colonialism, Resistance, and Reserves in British Columbia* (Vancouver: UBC Press, 2002), chap. 2. See also Paul Tennant, *Aboriginal People and Politics: The Indian Land Question in British Columbia, 1849-1989* (Vancouver: UBC Press, 1990), chap. 3; Chris Arnett, *The Terror of the Coast: Land Alienation and Colonial War on Vancouver Island and the Gulf Islands, 1849-1863* (Burnaby, BC: Talonbooks, 1999), chap. 2.

12 Hamar Foster and Alan Grove, "'Trespassers on the Soil': *United States v. Tom* and a New Perspective on the Short History of Treaty Making in Nineteenth-Century British Columbia," *BC Studies* 138-39 (2003): 51-84.

13 Douglas to Archibald Barclay, HBC Secretary, 3 September 1849, in Hartwell Bowsfield, ed., *Fort Victoria Letters: 1846-1852* (Winnipeg: Hudson's Bay Record Society, 1979), 43.

14 Barclay to Douglas, 17 December 1849, A.6/28 fos. 90d-92, Hudson's Bay Company Archives, Archives of Manitoba, Winnipeg.

15 Douglas to Barclay, 16 May 1850, in Bowsfield, ed., *Fort Victoria Letters*, 96 (emphasis added).

16 Barclay to Douglas, 16 August 1850, AC 20, Vi7 M430, BC Archives (BCA). On the New Zealand connection and understandings of Native title see Hamar Foster, "The Saanichton Bay Marina Case: Imperial Law, Colonial History and Competing Theories of Aboriginal Title," *UBC Law Review* 23 (1989): 629-50.

17 *Papers Connected with the Indian Land Question, 1850-1875, 1877*, a facsimile of a document printed in 1875, with a supplement published in 1877 (Victoria: Queen's Printers, 1987), 5-11 (emphasis added).

18 Wilson Duff, "The Fort Victoria Treaties," *BC Studies* 3 (1969): 4.

19 For a similar argument about understanding the treaties between the British and the Mi'kmaq see William C. Wicken, *Mi'kmaq Treaties on Trial: History, Land, and Donald Marshall Junior* (Toronto: University of Toronto Press, 2002).
20 On the possible meanings of "village sites" see C. Harris, *Making Native Space*, 25-26.
21 Douglas to House of Assembly, 5 February 1859, in *House of Assembly Correspondence Book, August 12th 1856 to July 6th 1859* (Victoria: W. Cullin, Printer to the King, 1918).
22 See Mackie, *Trading beyond the Mountains.*
23 See the discussion in Chapter 6. Also Douglas C. Harris, "The Boldt Decision in Canada: Aboriginal Treaty Rights to Fish on the Pacific," in *The Power of Promises: Perspectives on Treaties with Native Peoples in the Pacific Northwest,* ed. Alexandra Harmon (Seattle: University of Washington Press, 2008).
24 In correspondence over the application of the Reciprocity Treaty with the United States to the inhabitants of Vancouver Island, Herman Merivale (Permanent Under-Secretary, Colonial Office) wrote that the Crown grant of Vancouver Island to the HBC "not only omits the fisheries, but these were specifically and deliberately omitted." Merivale annotations to a letter from E. Hammond to Merivale, 13 June 1855, Colonial Office Correspondence, CO 305/6, p. 237, BCA.
25 *Prospectus for the Colonization of Vancouver Island* (London, 1849), emphasis added.
26 Sproat to I.W. Powell, 4 May 1878, in Federal Collection of Minutes of Decision, Correspondence and Sketches, vol. 1 (Letterbook 2, Joint Indian Reserve Commission and Sproat, March 1878 to January 1879), p. 64 (copy held by Department of Indian Affairs and Northern Development, Vancouver Regional Office).
27 Sproat to chief commissioner of Lands and Works, 4 May 1878, ibid., pp. 59-60.
28 Douglas to the Duke of Newcastle, 9 October 1860, Colonial Office Correspondence, CO 60/8, No. 11,678, BCA.
29 Sproat to E.A. Meredith, minister of the Interior, 30 July 1878, Department of Indian Affairs (DIA), RG 10, vol. 3662, file 9756, pt. 1, Library and Archives Canada (LAC). See the discussion in Chapter 2. See also Douglas C. Harris, *Fish, Law, and Colonialism: The Legal Capture of Salmon in British Columbia* (Toronto: University of Toronto Press, 2001), 136-40.
30 "Report of the Inspector of Fisheries for British Columbia," 3 January 1879, in Fisheries Annual Report 1878, Canada, *Sessional Papers,* 1879, p. 293.
31 "Report of the Fisheries Commissioner for 1918," p. X 11, British Columbia, *Sessional Papers,* 1919 (emphasis added).
32 See the discussion in Chapter 4.
33 See D.C. Harris, *Fish, Law, and Colonialism,* 18-27 and 61-65.
34 Lillian Ford and Cole Harris, "B.C. Colonial Indian Reserves" (unpublished paper, April 1999, archived at UBC Library, Special Collections).
35 C. Harris, *Making Native Space*, 30.
36 Although the Cowichan reserve along the lower reaches of the Cowichan River was considerably larger than other treaty or non-treaty reserves.
37 See the map in John Lutz, "Work, Wages and Welfare in Aboriginal-Non-Aboriginal Relations, British Columbia, 1849-1970" (PhD diss., Department of History, University of Ottawa, 1994), 149.
38 On the impact of the international border on the fisheries see Lissa Wadewitz, "The Nature of Borders: Salmon and Boundaries in the Puget Sound/Georgia Basin" (PhD diss., Department of History, University of California Los Angeles, 2004).
39 Dave Elliot Sr., *Salt Water People* (Saanich, BC: School District No. 63 (Saanich), 1983), 57.

40 Trutch's reductions of the Kamloops, Okanagan, and Stó:lō reserves are mapped in C. Harris, *Making Native Space*, 37-42, 56-61. For an analysis of Joseph Trutch's impact on colonial land policy see Robin Fisher, "Joseph Trutch and Indian Land Policy," *BC Studies* 12 (1971-72): 3-33.

41 *An Act to Amend an Act for the Preservation of Game* (1862), R.S.B.C. 1871, c. 12. See D.C. Harris, *Fish, Law, and Colonialism*, 36-39.

42 Douglas to the Duke of Newcastle, 31 August 1861, acknowledging receipt of the legislation, GR-1486, Colonial Office, CO 60/10, pp. 366-67, BCA.

43 *British North America Act,* 1867, ss. 91(12) and 91(24).

44 *Fisheries Act,* S.C., 1868, c. 60. The act was extended to British Columbia in May 1874 (S.C., 1874, c. 28) and proclaimed into force 1 July 1877 (Order-in-Council, May 1876, *Canada Gazette,* vol. 9, p. 1513).

45 *Salmon Fishery Regulations for the Province of British Columbia,* Order-in-Council, 30 May 1878, *Canada Gazette,* vol. 11, p. 1258.

46 J. Michael Thoms, "Ojibwa Fishing Grounds: A History of Ontario Fisheries Law, Science, and the Sportsmen's Challenge to Aboriginal Treaty Rights, 1650-1900" (PhD diss. Department of History, University of British Columbia, 2004).

47 Victor P. Lytwyn, "The Usurpation of Aboriginal Fishing Rights: A Study of the Saugeen Nation's Fishing Islands in Lake Huron," in *Co-Existence? Studies in Ontario-First Nations Relations,* ed. B.W. Hodgins, S. Heard, and J.S. Milloy (Peterborough, ON: Frost Centre for Canadian Heritage and Development Studies, 1992), 84-103.

48 Draper to J.M. Higginson, 16 April 1845, DIA, RG 10, vol. 612, p. 215, LAC. See Peggy J. Blair, "'For Our Race Is Our Licence': Culture, Courts and Conflict over Aboriginal Hunting and Fishing Rights in Southern Ontario" (LLD diss., University of Ottawa, 2003).

49 A view developed by Roland Wright, "The Public Right of Fishing, Government Fishing Policy, and Indian Fishing Rights in Upper Canada," *Ontario History* 86 (1994): 337.

50 Mark Walters, "Aboriginal Rights, *Magna Carta* and Exclusive Rights to Fisheries in the Waters of Upper Canada," *Queen's Law Journal* 23 (1988): 301-68.

51 Thoms, "Ojibwa Fishing Grounds," 244-47.

52 Canada (Province), *The Fishery Act* (1857), 20 Vict., c. 21.

53 Canada (Province), *The Fisheries Act* (1858), 22 Vict., c. 86, s. 4.

54 Canada (Province), *The Fisheries Act* (1865), 29 Vict., c. 11, s. 17(8). See Lise C. Hansen, "Treaty Rights and the Development of Fisheries Legislation in Ontario: A Primer," *Native Studies Review* 7 (1991): 1-21.

55 Article 13 of the Terms of Union says: "The charge of the Indians, and the trusteeship and management of the lands reserved for their use and benefit, shall be assumed by the Dominion Government, and a policy as liberal as that hitherto pursued by the British Columbia Government shall be continued by the Dominion Government after the Union." British Columbia, *Sessional Papers,* 1871, 12. In *R. v. Jack* (1979), [1980] 1 S.C.R. 294, [1979] 2 C.N.L.R. 25, Dickson J. found that this clause, in combination with the pre-Confederation colonial policy to recognize the priority of the Aboriginal fishery, required that the Dominion continue to recognize the priority of Aboriginal fishing. However, he concurred with the majority in the result because he found that the Department of Fisheries was justified in limiting the priority of the Native fishery for conservation purposes. In earlier work I mistakenly labelled Dickson J.'s decision "in dissent," when in fact he concurred in the result although for very different reasons. See Douglas C. Harris, "Indian Reserves, Aboriginal Fisheries, and the Public Right to Fish," in *Despotic Dominion: Property Rights in British Settler Colonies,* ed. John McLaren, A.R. Buck, and Nancy E. Wright (Vancouver: UBC Press, 2004), 266-93.

Chapter 2: Land Follows Fish

1 See the Department of Marine and Fisheries annual reports in the *Sessional Papers* of the Canadian Parliament through the 1870s.

2 For yearly cannery production in British Columbia see Cicely Lyons, *Salmon: Our Heritage* (Vancouver: British Columbia Packers, 1969), app. 36.

3 On Native participation in the industrial commercial fishery see Chapter 7.

4 See the population tables in Jean Barman, *The West beyond the West: A History of British Columbia,* rev. ed. (Toronto: University of Toronto Press, 1996), 379.

5 Cole Harris and Robert Galois, "A Population Geography of British Columbia in 1881," in *The Resettlement of British Columbia: Essays on Colonialism and Geographical Change,* by Cole Harris (Vancouver: UBC Press, 1997), 141.

6 See chap. 2 and the tables of pre-emption records (sorted by year and region) in app. B of Robert E. Cail, *Land, Man, and the Law: The Disposal of Crown Lands in British Columbia, 1871-1913* (Vancouver: UBC Press, 1974).

7 There are many accounts of these events, but for a focus on law see Tina Loo, *Making Law, Order, and Authority in British Columbia, 1821-1871* (Toronto: University of Toronto Press, 1994), 134-56.

8 Nicholas K. Blomley, "Law, Property, and the Geography of Violence: The Frontier, the Survey, and the Grid," *Annals of the Association of American Geographers* 93 (2003): 128-29.

9 *Papers Connected with the Indian Land Question, 1850-1875, 1877,* a facsimile of a document printed in 1875, with a supplement published in 1877 (Victoria: Queen's Printers, 1987).

10 See Hamar Foster, "Letting Go the Bone: The Idea of Indian Title in British Columbia, 1849-1927," in *Essays in the History of Canadian Law,* vol. 6, *British Columbia and the Yukon,* ed. Hamar Foster and John McLaren (Toronto: Osgoode Society for Canadian Legal History, 1995), 28-86; Cole Harris, *Making Native Space: Colonialism, Resistance, and Reserves in British Columbia* (Vancouver: UBC Press, 2002), 94-98.

11 Laird to Anderson, Memorandum of Instructions to the Dominion Commissioner, 25 August 1876, Department of Indian Affairs (DIA), RG 10, vol. 3633, file 6425-1 (reel C-10111), Library and Archives Canada (LAC) (emphasis added).

12 *Papers Connected with the Indian Land Question,* 130-31. D. Laird Memorandum, 1 March 1874, adopted by Dominion Order-in-Council, 24 April 1874 (emphasis added).

13 G.A. Walkem, "Report of the Government of British Columbia on the Subject of Indian Reserves," 17 August 1875, in *Papers Connected with the Indian Land Question* (emphasis added).

14 C. Good, deputy provincial secretary, to G.M. Sproat, Memorandum of Instructions, 23 October 1876, DIA, RG 10, vol. 3633, file 6425-1 (reel C-10111), LAC (emphasis added).

15 *R. v. Nikal,* [1996] 1 S.C.R. 1013, [1996] 3 C.N.L.R. 178; *R. v. Lewis,* [1996] 1 S.C.R. 921, [1996] 3 C.N.L.R. 131.

16 In his interpretation of the fisheries clauses in the mid-nineteenth-century treaties in the Washington Territory, Justice Boldt provided the following definitions for "fishing stations" and "fishing grounds": "'Stations' indicates fixed locations such as the site of a fish weir or a fishing platform or some other narrowly limited area; 'grounds' indicates larger areas which may contain numerous stations and other unspecified locations." *United States v. State of Washington* 384 F. Supp. 312, para. 8.

17 "Report of the Superintendent of Fisheries for Upper Canada," app. 30 of Annual Report of the Department of Fisheries, Canada, *Sessional Papers,* 1860, p. 79 (emphasis added).

18 Peggy Blair, in responding to the Supreme Court of Canada's distinction between land and fisheries, points out that in Ontario the Crown issued licences of occupation to

those entitled to use particular fishing stations, and that these licences included a description of the water boundaries of the exclusive fisheries. See Peggy J. Blair, "Settling the Fisheries: Pre-Confederation Crown Policy in Upper Canada and the Supreme Court's Decisions in *R. v. Nikal* and *Lewis,*" *Revue générale de droit* 31 (2001): 147.

19 *Fisheries Act,* S.B.C. 1901, c. 25, s. 3(8).
20 Regarding Lake Huron see Victor P. Lytwyn, "The Usurpation of Aboriginal Fishing Rights: A Study of the Saugeen Nation's Fishing Islands in Lake Huron," in *Co-Existence? Studies in Ontario-First Nations Relations,* ed. B.W. Hodgins, S. Heard, and J.S. Milloy (Peterborough, ON: Frost Centre for Canadian Heritage and Development Studies, 1992), 86.
21 New technology changed patterns of ownership within Native communities as well. Stó:lō fishers, for example, now use aluminum boats in their fisheries in the lower Fraser Canyon instead of land-based technology such as dip nets, and as a result they emphasize ownership of river eddies instead of particular points of land. See Keith Thor Carlson, "History Wars: Considering Contemporary Fishing Site Disputes," in *A Stó:lō -Coast Salish Historical Atlas,* ed. Keith Thor Carlson (Vancouver: Douglas and McIntyre, 2001), Plate 19.
22 I.W. Powell to superintendent general, Indian Affairs, 19 September 1877, DIA, RG 10, vol. 3651, file 8540 (reel C-10114), LAC (emphasis added).
23 On the conflict over the Cowichan weirs see Douglas C. Harris, *Fish, Law, and Colonialism: The Legal Capture of Salmon in British Columbia* (Toronto: University of Toronto Press, 2001), chap. 3. Reserve boundaries and fisheries on the Cowichan River were contested in *R. v. Jimmy* (1987), B.C.L.R. (2d) 145 (B.C.C.A.).
24 Sproat, report to the provincial government, 4 February 1878, DIA, RG 10, vol. 3657, file 9361 (reel C-10115), LAC.
25 On 4 January 1860, Governor Douglas issued the *Proclamation Relating to the Acquisition of Land* (Appendix to the R.S.B.C., 1871, c. 15) in the mainland colony of British Columbia, the first section of which allowed for the non-Native acquisition of land but excluded Native settlements: "From and after the date hereof, British subjects and aliens who shall take the oath of allegiance to Her Majesty and Her successors, may acquire unoccupied and unreserved and unsurveyed Crown Lands in British Columbia (not being the site of an existent or proposed town, or auriferous land available for mining purposes, or an *Indian Reserve or settlement,* in fee simple) under the following conditions ... " (emphasis added).
26 C. Harris, *Making Native Space,* 124-29.
27 Anderson to minister of Marine and Fisheries, 3 January 1878, DIA, RG 10, vol. 3651, file 8540 (reel C-10114), LAC.
28 Sproat to Indian Affairs, 4 February 1878, DIA, RG 10, vol. 3657, file 9361 (reel C-10115), LAC.
29 Edwards to C.B. Sword, 18 August 1906, Department of Marine and Fisheries, RG 23, file 3196, pt. 1, 12-13 (UBC reel 63), LAC.
30 The Department of Fisheries was not impressed. The inspector of Fisheries instructed Edwards to proceed with charges and to hire a lawyer if necessary to minimize the possibility that the charge would be dismissed. Sword to R.N. Venning, 30 August 1906, ibid., 14.
31 C. Harris, *Making Native Space,* 134. On the question of when Indian reserves became reserves in law see Hamar Foster, "Roadblocks and Legal History, Part 1: Do Forgotten Cases Make Good Law?" *The Advocate* 54 (1996): 355-66.
32 C. Harris, in *Making Native Space,* 140-44, describes Sproat's difficulties at Lytton and his painstaking efforts to find workable solutions. The best agricultural land on either side of Nahomin Creek, just south of the Stein River and formerly the site of a large

Nlha7kapmx settlement, had been pre-empted or purchased by settlers who had also recorded their rights to the water in the creek itself. Sproat allotted the land between these two properties – a boulder-strewn field that was all but unfarmable, particularly without water rights – as the Nahomin reserve and created a large temporary reserve to prevent further land sales or pre-emptions. He thought the latter might form the basis of a future resolution. The Dominion and provincial governments never recognized these temporary reserves.

33 For a fuller description see Keith Thor Carlson, "Innovation, Tradition, Colonialism, and Aboriginal Fishing Conflicts in the Lower Fraser Canyon," in *New Histories for Old: Changing Perspectives on Canada's Native Pasts,* ed. Ted Binnema and Susan Neylan (Vancouver: UBC Press, 2007), 145-74.

34 Field Minute, 1 June 1878, Federal Collection of Minutes of Decision, Correspondence and Sketches (Fed. Col.), vol. 4/1 (Field Minutes, Gilbert Malcolm Sproat, 1878-1880), pp. 12-13 (copy held by DIA, Vancouver Regional Office).

35 Sproat to Mancell, 25 May 1878, Fed. Col., vol. 1 (Letterbook 2, JIRC and Sproat, March 1878 to January 1879), pp. 123-25.

36 Matthew D. Evenden, *Fish versus Power: An Environmental History of the Fraser River* (Cambridge: Cambridge University Press, 2004), 19-52.

37 Field Minute, 1 June 1878, Fed. Col., vol. 4/1, p. 11.

38 Field Minute, 18 June 1878, Fed. Col., vol. 6 (G.M. Sproat, May 1878 to August 1878), p. 65.

39 Sproat to E. Mohun, 12 June 1878, Fed. Col., vol. 1. See also Mohun to Jennett, 1 July 1878, ibid.

40 In the same letter, Sproat raised, but offered no remedy for, the looming conflict between farmers, who held water rights to irrigate their fields, and the Kamloops, whose fishing stations would be worthless if there were insufficient water, after irrigation, to support fish.

41 Thanks to Anne Seymour for her help in documenting the fishery at Kamloops Indian Reserve No. 2 and the Bartlett Newman pre-emption.

42 *Salmon Fishery Regulations for the Province of British Columbia,* Order-in-Council, 30 May 1878, *Canada Gazette,* vol. 11, p. 1258.

43 Sproat to E.A. Meredith, minister of the Interior, 30 July 1878, DIA, RG 10, vol. 3662, file 9756, pt. 1 (reel C-10116), LAC.

44 W.F. Whitcher, commissioner of Fisheries, to Meredith, 5 August 1878, ibid. Whitcher had written to Meredith earlier, on 15 June 1878, to exempt Indians from the *Fisheries Act:* "It should be clearly understood that Indians and whitemen are all alike subject to the fishery laws. There is, however, no present intention to apply the Fisheries Act to the Indians of British Columbia" (ibid.). For a fuller discussion see D.C. Harris, *Fish, Law, and Colonialism,* 44-49.

45 Field Minute, 5 September 1878, Fed. Col., vol. 4 (Field Minutes, G.M. Sproat, 1878-1880), p. 10.

46 Sproat to Guichon, 14 September 1878, Fed. Col., vol. 1, p. 258.

47 Sproat to O'Keefe and Greenhow, 22 October 1878, ibid., p. 287.

48 House of Commons, *Official Report of Debates,* vol. 8 (21 April 1880), pp. 1633-34. *The Inland Sentinel,* 16 September 1880. See the entry for Barnard in the *Dictionary of Canadian Biography,* vol. 12 (Toronto: University of Toronto Press, 1990), 50-51.

49 A dozen years later, Barnard raised the issue in parliament again, backed by vigorous support from many quarters, including the editors of *The Inland Sentinel,* then based in Kamloops. See *The Inland Sentinel,* 9 April 1892, including the editorial and reproduction of the House of Commons debates. For earlier and continuing coverage see also 2 April 1892 and 9, 23, and 30 July 1892.

50 Sproat, 8 April 1880, Fed. Col., vol. 3 (JIRC and G.M. Sproat, "British Columbia Indian Reserves, Files Nos. 7571, 8132, 8402, 8496, 9804 and 20242 No. 1 Vol. No. 1" [Reg. No. B-64656]), p. 398 (emphasis added).

51 Sproat to superintendent general, Indian Affairs, 26 October 1878, Fed. Col., vol. 1, pp. 334-46.

52 Sproat to Meredith, 30 July 1878, ibid., pp. 193-97.

53 Sproat to deputy minister of the Interior, 6 November 1878, ibid., pp. 329-33.

54 Foster, "Letting Go the Bone," 31-34, 61-65.

55 C. Harris, *Making Native Space,* 159-61. On Sproat's commitment to Native self-government see Douglas C. Harris, "The Nlha7kapmx Meeting at Lytton, 1879, and the Rule of Law," *BC Studies* 108 (1995-96): 5-25.

56 Notes on Minutes of Decision, Yale, 5 August 1879, Fed. Col., vol. 5 (Interrupted Work Book No. 1, Gilbert Malcolm Sproat, circa June 1879 to June 1880), p. 21.

57 Minutes of Decision, Yale, 5 August 1879, ibid., p. 8.

58 Minutes of Decision, Union Bar and Hope, 12 and 16 August 1879, Fed. Col., vol. 18 (Gilbert Malcolm Sproat, May 1878 to January 1880 [Reg. No. B-64649]), pp. 207-8, 214.

59 See Carlson, "Innovation, Tradition, Colonialism."

60 The letter, signed by Chiefs Benedict, Joe Brown, Tom Peter, Billy Ernest, Charley Spintlum, Samuel Mack, and Jimmy Joe, was reported in the *Vancouver Province,* 7 March 1918.

61 Douglas to the Duke of Newcastle, 9 October 1860, Colonial Office Correspondence, CO 60/8, No. 11,678, BC Archives. See the discussion in Chapter 1. See also Daniel P. Marshall, "Claiming the Land: Indians, Goldseekers, and the Rush to British Columbia" (PhD diss., Department of History, University of British Columbia, 2000), 280.

62 Field Minutes, Spuzzum, 1 June 1880, Fed. Col., vol. 4/1, pp. 11-12.

Chapter 3: Exclusive Fisheries

 1 See the entry on O'Reilly by D.R. Williams in the *Dictionary of Canadian Biography,* vol. 13 (Toronto: University of Toronto Press, 1966), 788; Kenneth Brealey, "Travels from Point Ellice: Peter O'Reilly and the Indian Reserve System in British Columbia," *BC Studies* 115-16 (1997-98): 181-236.

 2 L. Vankoughnet to O'Reilly, 9 August 1880, Department of Indian Affairs (DIA), RG 10, vol. 3716, file 22195 (reel C-10125), Library and Archives Canada (LAC) (emphasis added).

 3 Michael Kew, "Salmon Availability, Technology, and Adaptation," in *A Complex Culture of the British Columbia Plateau: Traditional* Stl'átl'imx *Resource Use,* ed. Brian Hayden (Vancouver: UBC Press, 1992), 179.

 4 See Chapter 2.

 5 Steven Romanoff, "Fraser Lillooet Salmon Fishing," in Hayden, *A Complex Culture,* 470-505.

 6 Dorothy I.D. Kennedy and Randy Bouchard, "*Stl'átl'imx* (Fraser River Lillooet) Fishing," in Hayden, *A Complex Culture,* 266-354.

 7 Brealey, "Travels from Point Ellice," 199-201; Cole Harris, *Making Native Space: Colonialism, Resistance, and Reserves in British Columbia* (Vancouver: UBC Press, 2002), 172-74.

 8 O'Reilly, Minute of Decision, 4 July 1881, Federal Collection of Minutes of Decision, Correspondence and Sketches (Fed. Col.), vol. 8 (P. O'Reilly [Indian Reserve Commissioner], May 1881 to January 1882, file 29858, vol. 2 [Reg. No. B-64643]), p. 215b (copy held by DIA, Vancouver Regional Office).

 9 O'Reilly, Minute of Decision, 15 July 1881, ibid., p. 165.

10 O'Reilly, Minute of Decision, 21 July 1881, ibid., p. 103.

11 O'Reilly, Minute of Decision, 25 July 1881, ibid., p. 7; and Minute of Decision, 30 July 1881, Fed. Col., vol. 9, p. 303.

12 O'Reilly's recognition of an exclusive right to fish for the Fountain band formed the basis of defence counsel's unsuccessful argument in *R. v. Bob*, [1979] 4 C.N.L.R. 71 (B.C.C.C.), that the accused, a member of the Fountain band, did not need to respect a DFO closure of the salmon fishery in these waters.

13 Almost all of this reserve was surrendered in the 1950s to allow BC Electric to complete its Seton Lake hydroelectric project.

14 Kennedy and Bouchard, "*Stl'átl'imx* (Fraser River Lillooet) Fishing," 310-13.

15 Sebasha, QueeKwow, Guishood, Abraham, Ooyaa, Andamman, Memorial of the Chiefs of the Upper Nass Villages to I.W. Powell, 1 August 1881, DIA, RG 10, vol. 3766, file 32876 (reel C-10135), LAC.

16 Powell to superintendent general, Indian Affairs, 21 September 1881, DIA, RG 10, vol. 3802, file 50341 (reel C-10140), LAC.

17 Powell to O'Reilly, 7 May 1882, Fed. Col., vol. 7 (Correspondence to/from P. O'Reilly, Dr. I.W. Powell, et al., May 1881 to February 1884 [Reg. No. B-64643]), p. 104.

18 L. Vankoughnet to W.F. Whitcher, 17 October 1881, DIA, RG 10, vol. 3802, file 50341 (reel C-10140), LAC.

19 Moses McDonald, Paul Scow-gate, Joseph Morrison, Albert E. Nelson, James Hayward, Richard Wilson, Adam Clark, and Thomas Wright to O'Reilly, 5 October 1881, Fed. Col., vol. 7, p. 45.

20 Field Minutes, 25 March 1882, Fed. Col., vol. 9 (P. O'Reilly [Indian Reserve Commissioner], January 1882 to July 1882, file 29858, vol.3 [Reg. No. B-64644 (pages 1-130); Reg. No. B-64642 (pp. 149-294)]), p. 9.

21 *Letter from the Methodist Missionary Society to the Superintendent-General of Indian Affairs Respecting British Columbia Troubles* (Toronto: Methodist Church of Canada, 1889).

22 Thanks to Susan Marsden for help identifying Ts'ibasaa.

23 Contract between Paul Sebassah and C.S. Windsor, 23 June 1877, archived at UBC Library, Special Collections.

24 A decade earlier, Gilbert Malcolm Sproat had purchased land for his lumber mill from the colonial government and from the Nuu-chah-nulth. See Gilbert Malcolm Sproat, *The Nootka: Scenes and Studies of Savage Life*, ed. and annotated by Charles Lillard (1868; Victoria: Sono Nis Press, 1987), 4.

25 Contract between Paul Sebassah and the North West Commercial Company, 14 June 1878, archived at UBC Library, Special Collections. The following text is pasted to the bottom of the contract. It suggests that there was a secondary market for fishing rights secured under contract from Native peoples:

> *Extract from H.B.Co. records.*
> The "Otter" landed at Woodcocks Landing September 1875 and had on board Colonel Lane, who, the following spring purchased Woodcock's rights and built the first cannery on the Skeena at that point, renaming it Inverness.
> (Signed) T.S. French.

26 Peggy J. Blair, "Settling the Fisheries: Pre-Confederation Crown Policy in Upper Canada and the Supreme Court's Decisions in *R. v. Nikal* and *Lewis*," *Revue générale de droit* 31 (2001): 133-44; Blair, "Solemn Promises and *Solum* Rights: The Saugeen Ojibway Fishing Grounds and *R. v. Jones* and *Nadjiwon*," *Ottawa Law Review* 28 (1996-97): 125-43.

27 Victor P. Lytwyn, "The Usurpation of Aboriginal Fishing Rights: A Study of the Saugeen Nation's Fishing Islands in Lake Huron," in *Co-Existence? Studies in Ontario-First Nations Relations*, ed. B.W. Hodgins, S. Heard, and J.S. Milloy (Peterborough, ON: Frost Centre for Canadian Heritage and Development Studies, 1992), 84-103.

28 O'Reilly, Minute of Decision, 29 October 1881, Fed. Col., vol. 9, p. 45.
29 O'Reilly, Minutes of Decision, 20 October 1881, ibid., pp. 102-8.
30 In 1889, Fisheries would define the tidal boundary on the Nass River as "a line drawn across said River at right angles from a place known as Rocky Point, on the right bank thereof, immediately above Fishery Bay." Departmental Order, 28 September 1889, *Canada Gazette*, vol. 23, p. 546.
31 Field Minutes, Fed. Col., vol. 9, p. 32 (emphasis in original).
32 Brealey, "Travels from Point Ellice," 212-14.
33 Statement of Rev. T. Crosby, 22 March 1889, in "Appendix," *Letter from the Methodist Missionary Society*, p. 3.
34 For transcripts of the meetings, held 3 and 8 February 1887, see "Report of Conferences between the Provincial Government and Indian Delegations from Fort Simpson and Naas [sic] River," British Columbia, *Sessional Papers*, 1887, pp. 251-72.
35 Ibid., p. 257 (emphasis added).
36 Ibid., p. 259.

Chapter 4: Exclusive Fisheries and the Public Right to Fish
 1 William Blackstone, *Commentaries on the Laws of England*, a facsimile of the first edition of 1765-69 (Chicago: University of Chicago Press, 1979), 1:105.
 2 J.E. Côté, "The Reception of English Law," *Alberta Law Review* 15 (1977): 29-92.
 3 *Proclamation Having the Force of Law to Declare That English Law Is in Force in British Columbia*, 19 November 1858.
 4 William Blackstone, *Commentaries on the Laws of England*, ed. Wayne Morrison (London: Cavendish, 2001), 79. This publication presents the text of the ninth edition of the *Commentaries*. The passage quoted appears on page 107 of the original ninth edition.
 5 *The English Law Ordinance*, Appendix, R.S.B.C. 1871, no. 70.
 6 Bruce Ziff, "Warm Reception in a Cold Climate: English Property Law and the Suppression of the Canadian Legal Identity," in *Despotic Dominion: Property Rights in British Settler Societies*, ed. John McLaren, A.R. Buck, and Nancy E. Wright (Vancouver: UBC Press, 2005), 103-19.
 7 Mathew Hale, "De Juris Marie et Brachiorum Ejusdem," in *A Collection of Tracts Relative to the Law of England*, vol. 1, part 1, ed. F. Hargrave (1787; Abington, UK: Professional Books, 1982), 11.
 8 Mark Walters, "Aboriginal Rights, *Magna Carta* and Exclusive Rights to Fisheries in the Waters of Upper Canada," *Queen's Law Journal* 23 (1988): 315-16.
 9 *Malcolmson v. O'Dea* (1863), 10 H.L. Cas. 591.
10 Whitcher to L. Vankoughnet, 9 December 1881, Department of Indian Affairs (DIA), RG 10, vol. 3766, file 32876 (reel C-10135), Library and Archives Canada (LAC). By "public waters," Whitcher appears to have meant tidal or navigable waters.
11 *The Fisheries Act*, S.C. 1868, c. 60, s. 2, provided that the minister of Marine and Fisheries could license or lease fisheries, where exclusive fisheries did not exist already, for a term not exceeding nine years. Longer terms required the approval of the Governor-in-Council.
12 Macdonald to A.W. McLean, minister of Fisheries, 20 December 1881, DIA, RG 10, vol. 3766, file 32876 (reel C-10135), LAC (emphasis added).
13 McLean to Macdonald, 30 January 1882, ibid.
14 On the violent confrontation in Ontario see Victor P. Lytwyn, "Ojibway and Ottawa Fisheries around Manitoulin Island: Historical and Geographical Perspectives on Aboriginal and Treaty Rights," *Native Studies Review* 6 (1990): 1-30.
15 Anderson to Whitcher, 5 December 1881, in Department of Marine and Fisheries Annual Report, Canada, *Sessional Papers*, 1882, 2nd supp., p. 209.

16 Whitcher to L. Vankoughnet, 9 January 1883, Federal Collection of Minutes of Decision, Correspondence and Sketches (Fed. Col.), vol. 10 (P. O'Reilly [Indian Reserve Commissioner], June 1882 to February 1885, file No. 29858, vol. 4, [Reg No. B-64645]), p. 135 (copy held by DIA, Vancouver Regional Office).
17 Indian Affairs to Whitcher, 12 January 1883, ibid., p. 133.
18 Whitcher to Vankoughnet, 15 January 1883, ibid., pp. 129-31.
19 Indian Affairs to Whitcher, 26 January 1883, ibid., p. 127.
20 Walters, "Aboriginal Rights, *Magna Carta*," 324-26; Roland Wright, "The Public Right of Fishing, Government Fishing Policy, and Indian Fishing Rights in Upper Canada," *Ontario History* 86 (1994): 338-40.
21 W.H. Draper, Attorney General of the Province of Canada, to J.M. Higginson, 16 April 1845, DIA, RG 10, vol. 612, p. 215, LAC; Adam Wilson, Solicitor General, to the commissioner of Crown Lands, 11 March 1863, DIA, RG 10, vol. 323 (reel C-1224), LAC; James Cockburn, Solicitor General of the Province of Canada, Memorandum, 8 March 1866, DIA, RG 10, vol. 323 (reel C-1224), LAC.
22 Wright, "The Public Right of Fishing."
23 J. Michael Thoms, "Ojibwa Fishing Grounds: A History of Ontario Fisheries Law, Science, and the Sportsmen's Challenge to Aboriginal Treaty Rights, 1650-1900" (PhD diss., Department of History, University of British Columbia, 2004).
24 Peggy J. Blair, "'For Our Race Is Our Licence': Culture, Courts and Conflict over Aboriginal Hunting and Fishing Rights in Southern Ontario" (LLD diss., University of Ottawa, 2003).
25 Whitcher, Circular to Fisheries Overseers, 17 December 1875, DIA, RG 10, vol. 1972, file 5530, LAC.
26 Vankoughnet to Macdonald, 27 February 1882, DIA, RG 10, vol. 3766, file 32876 (reel C-10135), LAC.
27 Vankoughnet to Macdonald, 26 June 1882; Macdonald to McLean, 25 October 1882, ibid.
28 There is one exception: When O'Reilly returned to the north coast in 1888, he included within his minutes of decision for the Nass Indians (Nisga'a) Reserve No. 16 the allocation of "the exclusive right of salmon fishing in the Kin a mas river the entire length of the reserve a distance of about half a mile." O'Reilly, 8 September 1888, Fed. Col., vol. 11 (P. O'Reilly [Indian Reserve Commissioner], June 1885-March 1889, file 29858, vol. 5, file 20802 [extracts], [Reg. No. B-64646]), p. 35.
29 *Malcolmson v. O'Dea* (1863), 10 H.L. Cas. 593.
30 *Canada v. Robertson* (1882), 6 S.C.R. 52 (*Robertson*).
31 *British North America Act* (1867), now the *Constitution Act, 1867* (U.K.), 30 and 31 Vict. c. 3, reprinted in R.S.C. 1985, App. 2, No. 5.
32 *Robertson*, 78 (emphasis in original).
33 *Dixon v. Snetsinger* (1873), 23 U.C.C.P. 235.
34 *Robertson*, 82-88. Gwynne J.'s use of US cases was not unusual. See the analysis of the Supreme Court of Canada's reliance on US cases in S.I. Bushnell, "The Use of American Cases," *University of New Brunswick Law Journal* 35 (1986): 157-81.
35 *Robertson*, 89 (emphasis added).
36 Ibid.
37 Ibid., 114-15.
38 Ibid., 131.
39 Ibid., 136.
40 Ibid., 131-32.
41 *In Re Provincial Fisheries*, [1895] 26 S.C.R. 444 at 527 *[Provincial Fisheries Reference]* (emphasis added). See also pages 531-32.

42   Ibid., 528. Wright, in "The Public Right of Fishing," 339, cites this passage as evidence that the public right to fish precluded Crown grants of exclusive fisheries in navigable, non-tidal waters, but this is a misreading of Strong C.J.'s judgment. Strong's statement, reproduced above, that the public right to fish did not apply to non-tidal waters could not have been clearer. The "invariable practice" of recognizing the public's right to fish applied only to Crown lands for which the Crown had not alienated exclusive fisheries.
43   Justice Girouard concurred. *Provincial Fisheries Reference,* 548.
44   Ibid., 533.
45   *A.G. Canada v. A.G. Ontario,* [1898] A.C. 700 at 709, (1898) 12 C.R.A.C. 48 (H.L.).
46   *A.G. B.C. v. A.G. Canada,* [1914] A.C., 153 at 171, (1913) 15 D.L.R. 308 (H.L.).
47   *Calder v. A.G. B.C.,* [1973] S.C.R. 313, (1973) 34 D.L.R. (3d) 145; *Delgamuukw v. British Columbia,* [1997] 3 S.C.R. 1010, [1998] 1 C.N.L.R. 14. See Walters, "Aboriginal Rights, *Magna Carta*"; Peggy J. Blair, "Settling the Fisheries: Pre-Confederation Crown Policy in Upper Canada and the Supreme Court's Decisions in *R. v. Nikal* and *Lewis," Revue générale de droit* 31 (2001): 87-172.

Chapter 5: Indian Reserves and Fisheries
1   Kenneth Brealey, "Travels from Point Ellice: Peter O'Reilly and the Indian Reserve System in British Columbia," *BC Studies* 115-16 (1997-98): 181-236, includes maps of O'Reilly's routes, 194-96, and many of his sketches; Cole Harris, *Making Native Space: Colonialism, Resistance, and Reserves in British Columbia* (Vancouver: UBC Press, 2002), 169-215; Anne Elisabeth Seymour, "Natives and Reserve Establishment in Nineteenth-Century British Columbia" (master's thesis, Department of History, University of British Columbia, 1995).
2   L. Vankoughnet to O'Reilly, 9 August 1880, Department of Indian Affairs (DIA), RG 10, vol. 3716, file 22195 (reel C-10125), Library and Archives Canada (LAC).
3   Field Minutes, 9 October 1882, Federal Collection of Minutes of Decision, Correspondence and Sketches (Fed. Col.), vol. 10 (P. O'Reilly [Indian Reserve Commissioner], June 1882 to February 1885, file 29858, vol. 4, [Reg No. B-64645]), p. 340 (copy held by DIA, Vancouver Regional Office).
4   Field Minutes, 25 October 1882, ibid., p. 250.
5   Field Minutes, 30 October 1882, ibid., pp. 235-36 (emphasis added).
6   US courts would interpret this language as guaranteeing American Indians 50 percent of the commercial harvest. On the connections between the US cases and treaty interpretation in Canada see Douglas C. Harris, "The Boldt Decision in Canada: Aboriginal Treaty Rights to Fish on the Pacific," in *The Power of Promises: Perspectives on Treaties with Native Peoples in the Pacific Northwest,* ed. Alexandra Harmon (Seattle: University of Washington Press, 2008).
7   Field Minutes, 1 November 1882, Fed. Col., vol. 10, p. 220.
8   George Robinson (on behalf of the Haida Chiefs) to I.W. Powell, 12 September 1883, Fed. Col., vol. 7 (Correspondence to/from P. O'Reilly, Dr. I.W. Powell, et al., May 1881 to February 1884 [Reg. No. B-64643]), pp. 148A-48B.
9   Field Minutes, Village Island (Mamalelequala-Que'qua'sot'enox), 26 October 1886, Fed. Col., vol. 11 (P. O'Reilly [Indian Reserve Commissioner], June 1885-March 1889, file 29858, vol. 5, file 20802 [extracts], [Reg. No. B-64646]), p. 266.
10   Field Minutes, 30 October 1886, ibid., pp. 277-82.
11   S. Bray, Memorandum, 27 December 1888; O'Reilly to Indian Affairs, 14 January 1889, ibid., pp. 72-75.

12 Mowat to J. Tilton, deputy minister of Fisheries, 29 May 1890, Fed. Col., vol. 12 (P. O'Reilly [Indian Reserve Commissioner], April 1889 to January 1892, file 29858, vol. 6 [Reg. No. B-64647]), p. 138.
13 *Fisheries Regulations for the Province of British Columbia*, Order-in-Council, 26 November 1888, *Canada Gazette*, vol. 22, p. 956. On the Indian food fishery see Chapter 6.
14 K. Morrison to Mowat, 21 August 1890, DIA, RG 10, vol. 3828, file 60,926 (reel C-10145), LAC.
15 J. Gaudin, Fisheries officer, 18 December 1894, Department of Marine and Fisheries (DMF), RG 23, file 1469, pt. 1 (UBC reel 39), LAC.
16 A.W. Vowell, Indian reserve commissioner, to G.W. Morrow, Indian agent, 7 April 1905, DIA, RG 10, vol. 1282 (reel C-13902), LAC.
17 Sedgewick to J. Tilton, deputy minister of Fisheries, 15 August 1890, DIA, RG 10, vol. 3828, file 60926 (reel C-10145), LAC.
18 Vankoughnet to E. Dewdney, minister of the Interior, 7 August 1891, ibid.
19 Tupper to Dewdney, 24 August 1891, ibid. See Douglas C. Harris, *Fish, Law, and Colonialism: The Legal Capture of Salmon in British Columbia* (Toronto: University of Toronto Press, 2001), 69-73. On Tupper as minister of Fisheries see Margaret Beattie Bogue, *Fishing the Great Lakes: An Environmental History, 1783-1933* (Madison: University of Wisconsin Press, 2000), 216-37.
20 Indian Affairs to O'Reilly, 2 September 1891, DIA, RG 10, vol. 3828, file 60926 (reel C-10145), LAC.
21 O'Reilly, 7 September 1891, DIA, RG 10, vol. 3571, file 126, pt. A (reel C-10101), LAC. The site referred to was probably that at the outflow of Nilkitkwa Lake, which the McKenna-McBride Royal Commission would set aside as IR 21 and 21A in 1916. See Appendix, Figure A3.
22 O'Reilly, 30 September 1891, DIA, RG 10, vol. 3571, file 126, pt. A (reel C-10101), LAC.
23 *Fishery Regulations for the Province of British Columbia*, s. 3a, Order-in-Council, 26 November 1888, *Canada Gazette*, vol. 22, p. 956. See the discussion of the fisheries acts and regulations in Chapter 6.
24 O'Reilly to J.D. McLean, secretary Indian Affairs, 23 December 1897, DIA, RG 10, vol. 3908, file 107297-1 (reel C-10159, C-10160), LAC.
25 McLean to Indian Agents, 31 December 1897, ibid.
26 H. Guillod to McLean, 21 January 1898; C. Todd to McLean, 22 January 1898, ibid.
27 List attached to a letter from A.W. Vowell, BC Indian superintendent, to Indian Affairs, 8 February 1898, ibid. See maps in the Appendix.
28 C. Harris, *Making Native Space*, 206.
29 S. Stewart, Memorandum, 17 March 1898, DIA, RG 10, vol. 3908, file 107297-1 (reel C-10159, C-10160), LAC. In *R v. Jack* (1979), [1980] 1 S.C.R. 294, [1979] 2 C.N.L.R. 25, defence counsel Douglas Sanders based his defence of Cowichan fishers charged with *Fisheries Act* offences on the Douglas Treaties and the Terms of Union, even though the Cowichan were not a party to one of the treaties.
30 E.E. Prince, Memorandum, 5 April 1898, DMF, RG 23, vol. 306, file 2445, part 1 (reel T-3991), LAC.
31 McLean, Memorandum, 7 June 1898, DIA, RG 10, vol. 3909, file 107297-3 (reel C-10160), LAC.
32 Brealey, "Travels from Point Ellice," 222-23.
33 On Vowell's work as reserve commissioner, including the locations of his reserves, see C. Harris, *Making Native Space*, 219-28.

34 Vowell to superintendent general, Indian Affairs, 3 December 1900, in Department of Indian Affairs Annual Report, 1900, Canada, *Sessional Papers,* 1901, p. 299.
35 Vowell to superintendent general, Indian Affairs, 24 October 1901, in Department of Indian Affairs Annual Report, 1901, Canada, *Sessional Papers,* 1902, p. 290.
36 For a fuller discussion of the events surrounding the Babine fish weirs and the Barricade Treaty see D.C. Harris, *Fish, Law, and Colonialism,* chap. 2.
37 Deasy to deputy superintendent general, Indian Affairs, 31 March 1913 in Department of Indian Affairs Annual Report, 1913, Canada, *Sessional Papers,* 1914, pp. 257-58.
38 See Brian E. Titley, *A Narrow Vision: Duncan Campbell Scott and the Administration of Indian Affairs in Canada* (Vancouver: UBC Press, 1986).

Chapter 6: Constructing an Indian Food Fishery

1 Douglas C. Harris, *Fish, Law, and Colonialism: The Legal Capture of Salmon in British Columbia* (Toronto: University of Toronto Press, 2001), chap. 1.
2 *An Act Respecting the Extension and Application of "The Fisheries Act," to and in the Provinces of British Columbia, Prince Edward Island and Manitoba,* S.C. 1874, c. 28, extended the *Fisheries Act,* S.C. 1868, c. 60, to British Columbia. The act was proclaimed into force by order-in-council in May 1876 (*Canada Gazette,* vol. 9, p. 1513), to take effect 1 July 1877.
3 *Fisheries Act,* S.C. 1868, c. 60, s. 13(8) (emphasis added).
4 J. Michael Thoms, "Ojibwa Fishing Grounds: A History of Ontario Fisheries Law, Science, and the Sportsmen's Challenge to Aboriginal Treaty Rights, 1650-1900" (PhD diss., Department of History, University of British Columbia, 2004). See the discussion in Chapter 1 of this book.
5 Canada (Province), *The Fishery Act* (1857), 20 Vict., c. 21.
6 See Victor P. Lytwyn, "The Usurpation of Aboriginal Fishing Rights: A Study of the Saugeen Nation's Fishing Islands in Lake Huron," in *Co-Existence? Studies in Ontario-First Nations Relations,* ed. B.W. Hodgins, S. Heard, and J.S. Milloy (Peterborough, ON: Frost Centre for Canadian Heritage and Development Studies, 1992), 84-103.
7 A.C. Anderson, "Report of the Inspector of Fisheries for the Province of British Columbia, 1878," 3 January 1879, in app. 1 of the Commissioner of Fisheries Report, Canada, *Sessional Papers,* 1879, pp. 301-2. Many of these fishers and shoreworkers were Native men and women, and their participation in the industrial commercial fishery is the focus of the following chapter.
8 *Salmon Fishery Regulations for the Province of British Columbia,* Order-in-Council, 30 May 1878, *Canada Gazette,* vol. 11, p. 1258.
9 Sproat to E.A. Meredith, minister of the Interior, 30 July 1878, Department of Indian Affairs (DIA), RG 10, vol. 3662, file 9756 (reel C-10116), pt. 1, Library and Archives Canada (LAC). See also Chapter 2.
10 W.F. Whitcher, Dominion commissioner of Fisheries, to Meredith, 5 August 1878, ibid.
11 *Fishery Regulations for the Province of British Columbia,* s. 1, Order-in-Council, 26 November 1888, *Canada Gazette,* vol. 22, p. 956.
12 *Trout Fishery,* Order-in-Council, 26 November 1888, *Canada Gazette,* vol. 22, p. 956.
13 See Chapter 7.
14 British Columbia Fishery Commission Report, 1892, in Canada, *Sessional Papers,* 1893, p. 393.
15 *Fishery Regulations for the Province of British Columbia,* ss. 1, 6, Order-in-Council, 3 March 1894, *Canada Gazette,* vol. 27, p. 1579 (emphasis added).
16 Tupper to E. Dewdney, minister of the Interior, 24 August 1891, DIA, RG 10, vol. 3828, file 60926 (reel C-10145), LAC.

17 D.C. Harris, *Fish, Law, and Colonialism,* chap. 1.
18 *Fishery Regulations for the Province of British Columbia,* Order-in-Council, 8 June 1908, *Canada Gazette,* vol. 41, p. 3209, and Order-in-Council, 12 March 1910, *Canada Gazette,* vol. 43, p. 3206.
19 Teit, memo, read by Andy Paull into the record of a meeting between D.C. Scott, W.E. Ditchburn, and the executive committee of the Allied Tribes of British Columbia, 7 August 1923, transcript of the meeting, pp. 147-48, DIA, RG 10, vol. 3821, file 59335, pt. 4A (reel C-10143), LAC.
20 Transcript of Royal Commission on Indian Affairs for the Province of British Columbia Meeting with Dominion and Provincial Fisheries Officials in Regard to Fishing Privileges of Indians in BC, held in Victoria on 23 December 1915, pp. 20-23, DIA, RG 10, vol. 3908, file 107297-2 (reel C-10160), LAC.
21 F.H. Cunningham, chief inspector of Fisheries, to W.A. Found, 26 May 1913, Department of Marine and Fisheries (DMF), RG 23, file 6, part 8 (UBC reel 5), LAC.
22 Order-in-Council, 9 February 1915, *Canada Gazette,* vol. 49, supp., 20 February 1915, p. 67.
23 Transcript of Royal Commission Meeting with Dominion and Provincial Fisheries Officials ... , pp. 13, 23.
24 Order-in-Council, P.C. 2539, 11 September 1917, Dept. of Justice, RG 2, Vol. 1178, LAC.
25 Order-in-Council, 11 September 1917, *Canada Gazette,* vol. 51, p. 925 (emphasis added).
26 See the account in Matthew D. Evenden, *Fish versus Power: An Environmental History of the Fraser River* (Cambridge: Cambridge University Press, 2004), 19-36.
27 Indian Agent Graham of the Lytton Agency, Evidence to the Royal Commission on Indian Affairs, 4 November 1914, DIA, RG 10, vol. 11024, file AH7, LAC.
28 Report of the Fisheries Commissioner for 1917, British Columbia, *Sessional Papers,* 1918, Q9.
29 *Vancouver Province,* 18 May 1918.
30 *Vancouver Sun,* 12 July 1922, 4 July 1922. See also Andrea Laforet and Annie York, *Spuzzum: Fraser Canyon Histories, 1808-1939* (Vancouver: UBC Press, 1998), 99-102.
31 Evenden, *Fish versus Power,* 43-48.
32 Transcript of a meeting between Scott, Ditchburn, and the executive committee of the Allied Tribes, pp. 136-37.
33 Transcript of Royal Commission Meeting with Dominion and Provincial Fisheries Officials ... , p. 7.
34 *B.C. Fisherman* 1, 1 (15 May 1930): 17-18. See also the letter in support from "A Northern Fisherman," *B.C. Fisherman* 1, 3 (15 July 1930): 7.
35 Correspondence file, DMF, RG 23, vol. 2208, file 10-1-31, part 1, Pacific Region Federal Records Centre.
36 For a brief account of Charlie's life see Oliver N. Wells, ed., *Squamish Legends by Chief August Jack Khahtsahlano and Domanic Charlie* (Vancouver: Charles Chamberlain and Frank T. Coan, 1966).
37 Lillian Ford and Cole Harris, "B.C. Colonial Indian Reserves" (unpublished research paper, April 1999, archived at UBC Library, Special Collections), 16-17. For an account of the Squamish economy in the late nineteenth and early twentieth centuries see Chris Roine, "The Squamish Aboriginal Economy, 1860-1940" (master's thesis, Department of History, Simon Fraser University, 1996).
38 This was the one cut-off recommended by the commissioners in what was then the New Westminster Agency. See Royal Commission on Indian Affairs for the Province of British Columbia, *Report of the Royal Commission on Indian Affairs for the Province of British Columbia,* vol. 3, "New Westminster Agency" (Victoria: Acme Press, 1916), 624.

39 Canada Order-in-Council 1917-2544.

40 More land would be lost with the building of Lions Gate Bridge at First Narrows in 1936 (Canada Order-in-Council 1936-1691), but the royal commission cut-off lands would be returned.

41 Royal Commission on Indian Affairs, *Report,* vol. 3, "New Westminster Agency," 655.

42 W.E. Ditchburn, chief inspector of Fisheries, to C.C. Perry, Indian agent, 15 October 1923, DIA, RG 10, vol. 10899, file 167/20-2, LAC.

43 Motherwell to Perry, 1 December 1924, ibid.

44 See E. Palmer Patterson II, "Andrew Paull and the Canadian Indian Resurgence" (PhD diss., University of Washington, 1962).

45 Perry to Motherwell, 9 December 1924, DIA, RG 10, vol. 10899, file 167/20-2, LAC.

46 Motherwell to Perry, 22 December 1924, ibid.

47 Sam August to Perry, 8 October 1925, ibid.

48 Chief George Harry to Perry, 24 October 1925, ibid.

49 Arthur C. Sutton, constable, to Perry, 26 October 1925, ibid.

50 Perry to Motherwell, 29 October 1925, ibid.

51 Ibid.

52 Motherwell to Perry, 2 September 1925; Perry to Motherwell, 2 September 1925; Ditchburn to Perry, 28 November 1925, ibid.

53 Motherwell to Perry, 13 October 1925, ibid.

54 Perry to Motherwell, 29 October 1925, ibid.

55 "Fine Sport for Fishermen Here," *Vancouver Sun,* 6 November 1925, p. 4. See the call to protect the Capilano River for sport fishing in Thomas Wilson Lambert, *Fishing in British Columbia* (London: H. Cox, 1907), 70. The impact of sport fishing on Native fisheries is beginning to come into clearer focus. For British Columbia, see J. Michael Thoms, "A Place Called Pennask: Fly-Fishing and Colonialism at a British Columbia Lake," *BC Studies* 133 (2002): 69-98; D.C. Harris, *Fish, Law, and Colonialism,* chap. 3, "The Law Runs through It: Weirs, Logs, Nets, and Fly-Fishing on the Cowichan River, 1877-1937"; for Ontario, see Thoms, "Ojibwa Fishing Grounds"; and for the Maritimes, see Bill Parenteau, "'Care, Control and Supervision': Native People in the Canadian Atlantic Salmon Fishery," *Canadian Historical Review* 79 (1998): 1-35.

56 The various hearings are reported in the *Vancouver Province,* 3, 4, and 12 November, and 15 December 1925, and in the *Vancouver Sun,* 3 and 12 November 1925.

57 J.D. McLean to Perry, 26 October 1925, DIA, RG 10, vol. 10899, file 167/20-2, LAC.

58 *Special Fishery Regulations for the Province of British Columbia,* Order-in-Council, 26 April 1922 (P.C. 895), *Canada Gazette,* 29 April 1922.

59 *Indian Act,* R.S.C. 1906, c. 81, s. 33. "No person, or Indian other than an Indian of the band, shall without the authority of the Superintendent General, reside or hunt upon, occupy or use any land or marsh, or reside upon or occupy any road, or allowance for road, running through any reserve belonging to or occupied by such band."

60 "Fishing Rights under Dispute: Court Rules Indians Have Special Privileges," *Vancouver Sun,* 12 November 1925; "Indians May Fish Reserves: Indian Agent Beats Fishery Department in Police Court," *Vancouver Province,* 12 November 1925.

61 E. Hyde to Perry, 23 November 1925, DIA, RG 10, vol. 10899, file 167/20-2, LAC.

62 It is not clear if the parties argued over whether the Capilano River, where it ran through the Capilano reserve, formed part of the reserve. In his judgment, Judge Cayley appeared to proceed on the assumption that it did. His six-page decision in *Rex v. Charley* is archived at RG 10, vol. 10899, file 167/20-2, LAC. See also Judges McInnes and Cayley Minute Books, 1917-32, GR-1845, vol. 4, p. 274, BC Archives.

63 *British Columbia Terms of Union,* R.S.C. 1985, app. 2, no. 10, term 13.

64 This is the argument that Indian Affairs had sketched out in the 1890s and that Doug Sanders would put to the Supreme Court of Canada on behalf of the Cowichan Tribes fifty-four years later in *R. v. Jack* (1979), [1980] 1 S.C.R. 294, [1979] 2 C.N.L.R. 25.

65 One of the ironies of Cayley's decision is that Andrew Paull, as a young man just out of high school, had worked for Cayley, then a barrister and solicitor in Vancouver, from approximately 1907 to 1910. See H.F. Dunlop, *Andy Paull: As I Knew Him and Understood His Times* (Vancouver: Standard Press, 1983), 39-44; Patterson, "Andrew Paull," 44-48.

66 *Indian Act,* R.S.C. 1906, c. 81, s. 4.

67 Perry to Indian Affairs, 28 January 1926, DIA, RG 10, vol. 10899, file 987/20-2, LAC.

68 Scott to Ditchburn, 28 January 1926, ibid. This approach is consistent with Scott's restricted view of Native rights and title. See Brian E. Titley, *A Narrow Vision: Duncan Campbell Scott and the Administration of Indian Affairs in Canada* (Vancouver: UBC Press, 1986).

69 Resolution of the Squamish Indian Council, 30 April 1926, DIA, RG 10, vol. 10899, file 987/20-2, LAC.

70 Perry to Rev. G.H. Raley, 4 November 1926, ibid.

71 Receipts dated 9 March, 22 April, and 18 August 1927, ibid.

72 J. Whitmore to Ditchburn, 27 August 1926, ibid.

73 Perry to Motherwell, 31 August 1926, ibid.

74 Paull to Perry, 11 October 1926, ibid.

75 For a study of traditional Squamish marine resource use see Dorothy I.D. Kennedy and Randy Bouchard, *Utilization of Fish, Beach Foods, and Marine Mammals by the Squamish Indian People of British Columbia* (Victoria: British Columbia Indian Language Project, 1976), archived at UBC Library, Special Collections.

76 Perry to Ditchburn, 1 September 1926, DIA, RG 10, vol. 10899, file 987/20-2, LAC.

77 T.T. Bartlett to Perry, 26 September 1926, ibid.

78 Motherwell to Ditchburn, 6 July 1926, ibid.

79 See Geoff Meggs, *Salmon: The Decline of the British Columbia Fishery* (Vancouver: Douglas and McIntyre, 1995), 54.

80 Christopher Harvey, review of D.C. Harris, *Fish, Law, and Colonialism,* in *The Advocate* 61 (2003): 125-27.

Chapter 7: Licensing the Commercial Salmon Fishery

1 On the spatial distribution of the canning industry see Edward N. Higginbottom, "The Changing Geography of the Salmon Canning Industry in British Columbia, 1870-1931" (master's thesis, Simon Fraser University, 1988); Dianne Newell, "Dispersal and Concentration: The Slowly Changing Spatial Pattern of the British Columbia Salmon Canning Industry," *Journal of Historical Geography* 14 (1988): 22-36; Newell, *Tangled Webs of History: Indians and the Law in Canada's Pacific Coast Fisheries* (Toronto: University of Toronto Press, 1993), 16-20. For general histories see Geoff Meggs, *Salmon: The Decline of the British Columbia Fishery* (Vancouver: Douglas and McIntyre, 1995); Joseph E. Forester and Anne D. Forester, *Fishing: British Columbia's Commercial Fishing History* (Saanichton, BC: Hancock House, 1975); Cicely Lyons, *Salmon: Our Heritage* (Vancouver: British Columbia Packers, 1969).

2 *British North America Act,* 1867, s. 91(12).

3 See Chapter 4.

4 On the Davis Plan see Meggs, *Salmon,* part 3; Newell, *Tangled Webs,* chap. 7; Dennis Brown, *Salmon Wars: The Battle for the West Coast Salmon Fishery* (Madeira Park, BC: Harbour Publishing, 2005), chap. 6.

5 The Metlakatla cannery shipped nearly 30,000 cases of canned salmon in the four years it operated on the north coast in the 1880s. Department of Fisheries Annual Report, 1889, Canada, *Sessional Papers,* 1890, p. 249.

6 Rolf Knight, *Indians at Work: An Informal History of Native Indian Labour in British Columbia, 1858-1930* (Vancouver: New Star Books, 1978; rev. ed., 1996); John Lutz, "Work, Wages and Welfare in Aboriginal-Non-Aboriginal Relations, British Columbia, 1849-1970" (PhD diss., Department of History, University of Ottawa, 1994); Leonna Sparrow, "Work Histories of a Coast Salish Couple" (master's thesis, University of British Columbia, 1976).

7 See Douglas Hay and Paul Craven, eds., *Masters, Servants, and Magistrates in Britain and the Empire, 1562-1955* (Durham: University of North Carolina Press, 2004).

8 In Edward Gibbon Wakefield's influential model of the 1830s, this meant fixing a price for land that was neither so cheap as to be affordable to everyone, nor so expensive as to be out of reach of the industrious labourer. See "A View of the Art of Colonization, with Present Reference to the British Empire; in Letters between a Statesman and a Colonist," in M.F. Floyd Prichard, ed., *The Collected Works of Edward Gibbon Wakefield* (Auckland: Collins, 1969). On Wakefield's influence in British Columbia see Richard Mackie, "The Colonization of Vancouver Island, 1849-1858," *BC Studies* 96 (1992-93): 3-40; Cole Harris, *Making Native Space: Colonialism, Resistance, and Reserves in British Columbia* (Vancouver: UBC Press, 2002), 5.

9 See W. Peter Ward, *White Canada Forever: Popular Attitudes and Public Policy towards Orientals in British Columbia,* 2nd ed. (Montreal and Kingston: McGill-Queen's University Press, 1990); Kay Anderson, *Vancouver's Chinatown: Racial Discourse in Canada, 1875-1980* (Montreal and Kingston: McGill-Queen's University Press, 1991).

10 See Alicja Muszynski, *Cheap Wage Labour: Race and Gender in the Fisheries of British Columbia* (Montreal and Kingston: McGill-Queen's University Press, 1996).

11 *Canada v. Robertson* (1882), 6. S.C.R. 52; *A.G. Canada v. A.G. Ontario,* [1898] A.C. 700, (1898) 12 C.R.A.C. 48.

12 *British Columbia Fisheries Act,* S.B.C. 1901, c. 25. The act was substantially amended in 1902, S.B.C. 1902, c. 26. It was not proclaimed into force until 1907 (British Columbia Order-in-Council, 4 June 1907). Earlier provincial regulation of the fisheries appeared in the *Game Protection Act,* S.B.C. 1888, c. 52. One fisher's refusal to purchase a provincial licence required under the provincial *Fisheries Act* led to the case of *North v. Kendall* (14 December 1904) (B.C.S.C.), transcript of oral decision in the BC Archives, GR-0429, box 16, file 2, and a ruling that the province had no jurisdiction to regulate the salmon fishing season.

13 *A.G. B.C. v. A.G. Canada,* [1914] A.C. 153 at 171, (1913) 15 D.L.R. 308 (H.L.).

14 Forester and Forester, *Fishing,* 55-59; Richard Rathburn, *A Review of the Fisheries in the Contiguous Waters of the State of Washington and British Columbia* (Washington, 1899), 305-10.

15 S. Wilmot, "Salmon Fishery and Fishery Regulations of Fraser River, B.C.," in Department of Fisheries Annual Report, 1890, Canada, *Sessional Papers,* 1891, pp. 65-66.

16 This licensing requirement followed the *Fishery Regulation,* Order-in-Council, 11 June 1879, which prohibited all salmon fishing except under Fisheries lease or licence.

17 George A. Rounsefell and George B. Kelez, *The Salmon and Salmon Fisheries of Swiftsure Bank, Puget Sound, and the Fraser River* (Washington, DC: US Government Printing Office, 1938), 706. The authors note that "from 1877 to 1899 the nationalities have been estimated from various notes." They do not identify the notes, and some of the figures seem a little high when compared to the figures that are available in the Department of Fisheries annual reports in the *Sessional Papers* for Canada. But if the specific numbers are not entirely accurate, Rounsefell and Kelez's table is useful for the trends in

the fishery that it reveals. See Keith Ralston's assessment, "The 1900 Strike of Fraser River Sockeye Salmon Fishermen" (master's thesis, Department of History, University of British Columbia, 1965), 11n24 and 48.

18 On the state of the Columbia River fishery see Joseph E. Taylor III, *Making Salmon: An Environmental History of the Northwest Fisheries Crisis* (Seattle: University of Washington Press, 1999), 39-67.

19 Under the *Fishery Regulations for the Province of British Columbia*, 1888, s. 5, the minister of Marine and Fisheries had the capacity to "determine the number of boats, seines, or nets, or other fishing apparatus to be used in any of the waters of British Columbia." Canada Order-in-Council, 26 November 1888, *Canada Gazette*, vol. 22, p. 956.

20 Fisheries Report, 1889, Canada, *Sessional Papers*, 1890, p. 254.

21 *Fishery Regulations for the Province of British Columbia*, Canada Order-in-Council, 14 March 1890, *Canada Gazette*, vol. 23, p. 1903. See Meggs, *Salmon*, chap. 3.

22 Rounsefell and Kelez, *Salmon and Salmon Fisheries*, 706.

23 British Columbia Fishery Commission, 1892, Record of Proceeding and Minutes of Evidence, in Canada, *Sessional Papers*, 1893, p. 392.

24 W.H. Lomas, Indian agent, to John McNab, inspector of Fisheries, 26 January 1892; McNab to S.P. Bauset, acting deputy minister, Department of Marine and Fisheries, 29 February 1892, Department of Indian Affairs (DIA), RG 10, vol. 3828, file 60926 (reel C-10145), Library and Archives Canada (LAC).

25 Douglas C. Harris, *Fish, Law, and Colonialism: The Legal Capture of Salmon in British Columbia* (Toronto: University of Toronto Press, 2001), 144-45.

26 Canada Order-in-Council, 3 March 1894, *Canada Gazette*, vol. 27, p. 1579.

27 Keith Ralston estimates that by 1900, only 12 percent of licences on the Fraser were held by canneries ("The 1900 Strike," 33).

28 Bell-Irving to Fisheries, 19 February 1912, GR 435, box 65, file 607, British Columbia Archives (BCA).

29 David Reid, "The Development of the Fraser River Salmon Canning Industry, 1885 to 1913" (report prepared for Fisheries and Marine Service, Pacific Region, Department of the Environment, July 1973), 84.

30 Canada Order-in-Council, 29 March 1899, *Canada Gazette*, vol. 32, p. 1884.

31 C.B. Sword, evidence to the Dominion-British Columbia Fisheries Commission, 1905-7, 17 November 1905, p. 157, archived at UBC Library, Special Collections.

32 Ralston, "The 1900 Strike."

33 Rounsefell and Kelez, *Salmon and Salmon Fisheries*, 706.

34 D.C. Scott, deputy superintendent general of Indian Affairs, to M. Burrell, 11 August 1915, DIA, RG 10, vol. 3909, file 107297-3 (reel C-10160), LAC.

35 J.A. Motherwell, chief inspector of Fisheries, to J.P. Babcock, 9 August 1922, GR 435, box 43, file 392, BCA. Motherwell outlines the goals for 1923 of a 15 percent reduction of the Japanese fleet in Districts 1 and 3, and a 10 to 50 percent reduction in District 2. Ken Adachi, *The Enemy That Never Was: A History of the Japanese Canadians* (Toronto: McClelland and Stewart, 1991), 142-45; Daphne Marlatt, ed., *Steveston Recollected: A Japanese-Canadian History* (Victoria: Provincial Archives of British Columbia, 1975), 53-55; Hozumi Yonemura, "Japanese Fishermen in British Columbia and British Fair Play," *Canadian Forum* 10 (1930): 357.

36 Daniel Boxberger, *To Fish in Common: The Ethnohistory of Lummi Indian Salmon Fishing* (Lincoln: University of Nebraska Press, 1989; Seattle: University of Washington Press, 2000), 35. See also Boxberger, "In and Out of the Labour Force: The Lummi Indians and the Development of the Commercial Salmon Fishery in North Puget Sound, 1880-1900," *Ethnohistory* 35 (1988): 161-90.

37 See the description of a reef-net fishery in Hilary Stewart, *Indian Fishing: Early Methods on the Northwest Coast* (Vancouver: Douglas and McIntyre, 1977), 93-94.
38 Testimony of Squamish chief Mathias Joseph, New Westminster Agency, 21 June 1915, Royal Commission on Indian Affairs for the Province of British Columbia, Evidence, vol. 6, pp. 40-41, archived at UBC Library, Special Collections.
39 Testimony of Musqueam chief Jonnie, 24 June 1913, ibid., p. 64.
40 Under the *Indian Act*, S.C. 1876, c. 18, s. 66, no person could take real or personal property held by an Indian and located on a reserve as security for a loan. The one exception was for the seller of an item who might reclaim that item to recover money owed for its purchase. In the fishery, this exception might have provided some security for a boat seller, but nets only lasted two seasons and quickly lost their value.
41 Unsigned copy of contract, 23 February 1910, GR 435, box 14, file 119, BCA.
42 John Charles Pritchard suggests a similar agreement existed among canners on Rivers Inlet from 1903. Pritchard, "Economic Development and Disintegration of Traditional Culture among the Haisla" (PhD diss., Department of Anthropology and Sociology, University of British Columbia, 1977), 75. For a splendid fictional account of collusion in the British Columbia canning industry see Bertrand W. Sinclair, *Poor Man's Rock* (Boston: Little, Brown, 1920).
43 The *General Fishery Regulations for the Province of British Columbia*, 1908, removed the limit on the number of licences that canneries could hold. Order-in-Council, 8 June 1908, *Canada Gazette*, vol. 41, p. 3209.
44 T.G. Wynn to J.P. Babcock, 2 August 1909, GR 435, box 70, file 659, BCA.
45 Canada Order-in-Council, 22 December 1910, *Canada Gazette*, vol. 44, p. 2170.
46 Memorandum from W.A. Found, Dominion superintendent of Fisheries, and D.N. McIntyre, provincial deputy commissioner of Fisheries, 7 August 1912, GR 435, box 16, file 137, BCA (emphasis added).
47 Cole Harris and Robert Galois, "A Population Geography of British Columbia in 1881," in *The Resettlement of British Columbia: Essays on Colonialism and Geographical Change*, by Cole Harris (Vancouver: UBC Press, 1997), 138-40, 149. For general population figures see Jean Barman, *The West beyond the West: A History of British Columbia*, rev. ed. (Toronto: University of Toronto Press, 1996), 379, 387.
48 Scottish Records Office, MS-1640, BCA.
49 A. Johnston to deputy commissioner of Fisheries for BC, 9 November 1912, Department of Marine and Fisheries (DMF), RG 23, file 6, pt. 7 (UBC reel 5), LAC (emphasis in original). See also British Columbia Order-in-Council 1404, 15 October 1913, an explanation of the policy in response to a complaint from the Japanese government.
50 J.D. Hazen, minister of Marine and Fisheries, to A.E. O'Meara, general counsel for the Allied Tribes, 21 November 1912, DMF, RG 23, file 6, pt. 7 (UBC reel 5), LAC.
51 Canners to W.J. Bowser, BC Attorney General, 27 November and 17 December 1912, GR 435, box 16, file 137, BCA.
52 Memorandum of Meeting of Delegation of BC Canners with J.D. Hazen, 24 February 1912, GR 435, box 65, file 607, BCA.
53 John Locke, "Of Property," in *Two Treatises of Government*, ed. Peter Laslett (Cambridge: Cambridge University Press, 1960), 327-44.
54 G.D. Tite and O.H. Nelson, president and secretary, Prince Rupert Conservative Association, to H.S. Clements, MP, DMF, RG 23, file 6, pt. 7 (UBC reel 5), LAC.
55 BC Packers to Hazen, 19 August 1913, DMF, RG 23, file 6, pt. 8 (UBC reel 5), LAC.
56 Muszynski, *Cheap Wage Labour.*
57 Clements to Hazen, 27 February 1913, DMF, RG 23, file 6, pt. 7 (UBC reel 5), LAC.
58 J.M. Macmillan, Cassiar Packing Co., to Hazen, 22 March 1913, ibid.
59 Skeena Canners to Hazen, 17 March 1913, ibid.

60 Bowser to Canadian Cold Storage, 16 April 1913, ibid. See British Columbia Order-in-Council 1404, 15 October 1913. The province, which had been attempting to assert its authority over the fisheries since 1901, when it passed the *Fisheries Act*, S.B.C. 1901, c. 25 (although not proclaimed into force until 1907, British Columbia Order-in-Council, 4 June 1907), conceded in 1916, after a series of unsuccessful court cases, that it had no authority to restrict the granting of licences and acquiesced to the Dominion's allocation of twenty-five licences to Canadian Fish and Cold Storage for Native fishers.

61 Annual Fisheries Report for year end 31 March 1916, British Columbia, *Sessional Papers*, 1916, p. 245.

62 Pritchard, "Economic Development," 77-79.

63 Special Fisheries Commission, *Report of Special Fisheries Commission, 1917* (Ottawa, 1918), 33.

64 The fur trade had produced an earlier reliance on a single activity, creating patterns of dependence when furs disappeared or markets collapsed. See Arthur J. Ray, "Periodic Shortages, Native Welfare, and the Hudson's Bay Company 1670-1930," in *The Subarctic Fur Trade: Native Social and Economic Adaptations*, ed. Shepherd Krech III (Vancouver: UBC Press, 1984), 1-20.

65 GR 435, box 17, file 153, BCA.

66 M. Johnson, R. Brown, and others to Hazen, 11 August 1913, DMF, RG 23, file 6, pt. 8 (UBC reel 5), LAC.

67 Port Simpson Petition for Independent Licenses to the Provincial Legislature, 3 December 1913, GR 435, box 17, file 149, BCA.

68 McIntyre to commissioner of Fisheries, 21 February 1914, ibid.

69 Royal Commission on Indian Affairs for the Province of British Columbia, Evidence, vol. 1, p. 423, archived at UBC Library, Special Collections.

70 Patmore and Fulton to A. Johnston, 5 March 1914; Johnston to Patmore and Fulton, 18 March 1914, DMF, RG 23, file 6, pt. 8 (UBC reel 5), LAC.

71 Testimony of Benjamin Bennet, Nass Agency, 25 September 1915, DIA, RG 10, vol. 11025, file AH8 (reel T-3963), LAC.

72 Royal Commission on Indian Affairs for the Province of British Columbia, *Confidential Report of the Royal Commission on Indian Affairs for the Province of British Columbia* (Victoria: Acme Press, 1916), 15. See the fuller discussion in Chapter 8.

73 Ibid., 17.

74 Evidence to the Special Fisheries Commission, 1917, p. 274, cited in Pritchard, "Economic Development," 79.

75 Special Fisheries Commission, *Report, 1917*, 35.

76 DIA, RG 10, series C-V-8, vol. 11147, file 18, pt. A, LAC.

77 "Memo Re New Policy of Fisheries Department with Regard to Issuing Licenses," 10 January 1920, GR 435, box 43, file 387, BCA.

78 Forester and Forester, *Fishing*, 64-66.

79 Resolutions passed by the Cowichan Council, 4 June 1888, DIA, RG 10, vol. 3081, file 49287, LAC.

80 Canada Order-in-Council, 28 September 1889, *Canada Gazette*, vol. 23, p. 546.

81 Canada Order-in-Council, 7 November 1890, *Canada Gazette*, vol. 14, p. 876.

82 *Fisheries Act*, R.S.C. 1886, c. 95, as amended by S.C. 1891 c. 43.

83 Department of Fisheries Annual Reports, 1890, 1892, 1897, in Canada, *Sessional Papers*, 1891, 1893, 1898.

84 *Fishery Regulations for the Province of British Columbia* (1894), ss. 5 and 8, Canada Order-in-Council, 3 March 1894, *Canada Gazette*, vol. 27, p. 1579.

85 Statements of J. Gaudin, R.B. McMicking, J. Rood, Mr. Aikman, and J.T. Walbran, 18 December 1894, DMF, RG 23, file 1469, pt. 1 (UBC reel 39), LAC.

86 Department of Marine and Fisheries Annual Report, in Canada, *Sessional Papers,* 1898, p. 226.
87 J. McNab to deputy minister of Fisheries, 8 November 1898, DMF, RG 23, file 583, pt. 1, LAC.
88 Chief Joe Komiaken to W.W. Stumbles, 28 October 1899, ibid.
89 F. Gourdeau to G. O'Reilly, MP, 19 May 1902, DMF, RG 23, file 678, pt. 4 (UBC Reel 28), LAC.
90 F. Pedley to Gourdeau, 16 February 1903, and Gourdeau to Pedley, 28 February 1903, DMF, RG 23, vol. 2136 (1), (1903), Pacific Region Federal Records Centre (PRFRC).
91 Fisheries to A. Haslam, MP, 27 June 1895, DMF, RG 23, file 678, pt. 1 (UBC reel 26), LAC.
92 See Rathburn, *Review of the Fisheries.*
93 See Lissa Wadewitz, "The Nature of Borders: Salmon and Boundaries in the Puget Sound/Georgia Basin" (PhD diss., Department of History, University of California Los Angeles, 2004), on the efforts of Native and non-Native fishers to raid and otherwise disrupt the salmon-trap fishery.
94 Vowell to J. Gaudin, 6 February 1900, DMF, RG 23, file 678, pt. 2 (UBC reel 27), LAC.
95 *Land Act,* R.S.B.C. 1888, c. 66, s. 53, as amended by S.B.C 1894, c. 24, s. 2.
96 British Columbia Order-in-Council, 27 March 1903.
97 See the notice in the *Vancouver Province,* 10 December 1904.
98 See British Columbia *Sessional Papers,* 1902, p. 1236. British Columbia Canning Co. to chief commissioner of Lands and Works, 17 October 1901; Fell and Gregory to deputy Attorney General, 1 November 1904, GR 429, box 11, file 5, BCA.
99 F.H. Cunningham, chief inspector of Fisheries, to D.N. McIntyre, provincial deputy commissioner of Fisheries, 8 February 1913, GR 435, box 16, file 139, BCA, reveals a considerable degree of cooperation between the two departments over the allocation of licences.
100 I.W. Powell to superintendent general of Indian Affairs, 28 October 1879, DIA, RG 10, vol. 3662, file 9756, pt. 1 (reel C-10116), LAC.
101 E.L. Newcombe, deputy minister of Justice, to Indian Affairs, 28 February 1905, DIA, RG 10, vol. 10240, file 901/30-1-13 (reel T-7543), LAC.
102 *Land Registry Act Amendment Act,* S.B.C. 1910, c. 27, s. 48(b).
103 R.T. Burtwell to E.E. Prince, 7 October 1904; R.N. Venning to Burtwell, 28 October 1904, DMF, RG 23, file 678, pt. 6 (UBC reel 28), LAC.
104 Statement of the Allied Tribes of British Columbia for the Government of British Columbia, 12 November 1919, DIA, RG 10, vol. 3821, file 59335, part 4A (reel C-10143), LAC. See also the memo from the counsel for the Allied Tribes, A.E. O'Meara, "Foreshores in Front of Indian Reserves," 23 August 1923, ibid. Vina A. Starr, "Indian Title to Foreshore on Coastal Reserves in British Columbia" (report prepared for Department of Indian Affairs and Northern Development, 28 March 1985).
105 The 1904 regulations prohibited salmon traps and seines "within three miles of the mouth of a navigable river, and within half [a] mile of the mouth of a salmon stream, or in any special locality named by the Department." *Fishery Regulations, 1894,* s. 5(12) as amended by Canada Order-in-Council, 2 May 1904, *Canada Gazette,* vol. 37, p. 2207. These restrictions diminished, so that by 1908 there were few additional restrictions for those using seine nets. Canada Order-in-Council, 8 June 1908, *Canada Gazette,* vol. 41, p. 3209.
106 Pedley to Gourdeau, 4 April 1905, DMF, RG 23, file 3023, pt. 1 (UBC reel 57), LAC.
107 Vowell to Secretary, Indian Affairs, 29 August 1905, DIA, RG 10, Vol. 1282, p. 161-62 (reel C-13902), LAC.

108 J.T. Williams to Venning, 16 May 1905, DMF, RG 23, file 3023, pt. 1 (UBC reel 57), LAC.
109 Royal Commission on Indian Affairs, Bella Coola Agency Evidence, 25 August 1913, DIA, RG 10, vol. 11024, file AH2 (reel T-3961, T-3962), LAC.
110 Venning to Williams, 4 October 1906, DMF, RG 23, file 3023, pt. 1 (UBC reel 57), LAC.
111 G.W. Morrow to Williams, 16 June 1906, ibid.
112 *An Act to Amend the Land Act,* S.B.C. 1907, c. 25, s. 9(13). "The right to apply for permission to purchase under this section shall not be held to extend to any of the aborigines of this continent, except to such as shall have obtained permission to so apply by a special order of the Lieutenant-Governor in Council." This apparently confirmed the provincial government's practice not to sell land to Natives.
113 G. Macaulay to Fisheries, 24 December 1911, DMF, RG 23, file 3023, pt. 3 (UBC reel 59), LAC.
114 Cunningham to Fisheries, 23 January 1912, ibid.
115 Perry to Indian Affairs, 19 August 1912, ibid.
116 Cunningham to W.A. Found, 21 October 1912, ibid.
117 Perry to Indian Affairs, 19 August 1912, ibid.
118 Cunningham to McIntyre, 20 August 1915, GR 435, box 41, file 367, BCA.
119 E.G. Taylor to W.M. Halliday, 2 November 1915, DIA, RG 10, series C-V-8, vol. 11147, file 18, pt. A, LAC.
120 Testimony of 'Namgis chief Lageuse, Kwawkewlth Agency, 2 June 1914, DIA, RG 10, vol. 11025, file AH6 (reel T-3962), pp. 142-43, LAC.
121 Transcript of a meeting between the executive committee of the Allied Tribes, other Native leaders, Deputy Superintendent General of Indian Affairs D.C. Scott and Chief Inspector of Fisheries W.E. Ditchburn, 7-8 August 1923, pp. 157-66, DIA, RG 10, vol. 3821, file 59335, part 4A (reel C-10143), LAC.
122 J.A. Motherwell to Ditchburn and others, 11 October 1923, DMF, RG 23, vol. 2228, file 10-5-31, PRFRC.
123 E. Pinkerton, "Indians in the Fishing Industry," in *Uncommon Property: The Fishing and Fish Processing Industries in British Columbia,* ed. P. Marchak, N. Guppy, and J. McCullan (Toronto: Methuen, 1987), 256; C. Harris, *Making Native Space,* 286-89. See Lutz, "Work, Wages and Welfare," 18-23, for a somewhat different view.
124 Meggs, *Salmon,* 48-50.
125 Muszynski, in *Cheap Wage Labour,* highlights the interplay of class, race, and gender in the BC fisheries in her effort to explain the hiring choices in the salmon industry and to rethink Marx's theory of value.
126 *R. v. Kapp,* [2003] 4 C.N.L.R. 238, para. 38.

Chapter 8: Land and Fisheries Detached

1 Robert Galois, "The Indian Rights Association, Native Protest Activity and the 'Land Question' in British Columbia, 1903-1916," *Native Studies Review* 8 (1992): 113-83; Paul Tennant, *Aboriginal People and Politics: The Indian Land Question in British Columbia, 1849-1989* (Vancouver: UBC Press, 1990), chap. 7; Keith Thor Carlson, "Rethinking Dialogue and History: The King's Promise and the 1906 Aboriginal Delegation to London," *Native Studies Review* 16, 2 (2005): 1-38.
2 On the legal opinion behind the proposed reference see Hamar Foster, "A Romance of the Lost: The Role of Tom MacInnes in the History of the British Columbia Indian Land Question," in *Essays in the History of Canadian Law,* vol. 8, *In Honour of R.C.B. Risk,* ed. G. Blaine Baker and Jim Phillips (Toronto: Osgoode Society for Canadian Legal History, 1999), 171.

3 The work of the McKenna-McBride Commission deserves a book-length study. On its formation and deliberations over land see Robert E. Cail, *Land, Man, and the Law: The Disposal of Crown Lands in British Columbia, 1871-1913* (Vancouver: UBC Press, 1974), 227-43; Cole Harris, *Making Native Space: Colonialism, Resistance, and Reserves in British Columbia* (Vancouver: UBC Press, 2002), 228-48.

4 McKenna to McBride, 29 July 1912, GR 441, box 147, file 1, British Columbia Archives (BCA).

5 Ibid.

6 Royal Commission on Indian Affairs, Cowichan Agency Evidence, Department of Indian Affairs (DIA) RG 10, vol. 11024, file AH3 (reel T-3962), Library and Archives Canada (LAC).

7 Ibid.

8 For an analysis of the testimony of coastal tribes see Deidre Sanders, Naneen Stuckey, Kathleen Mahoney, and Leland Donald, "What the People Said: Kwakwaka'wakw, Nuu-chah-nulth, and Tsimshian Testimonies before the Royal Commission on Indian Affairs for the Province of British Columbia (1913-1916)," *Canadian Journal of Native Studies* 19 (1999): 213-48.

9 Wetmore to Crothers, acting superintendent general of Indian Affairs, 9 June 1913, DIA, RG 10, vol. 11026, file FR1 (reel T-3964), LAC. See Douglas C. Harris, *Fish, Law, and Colonialism: The Legal Capture of Salmon in British Columbia* (Toronto: University of Toronto Press, 2001), 168-72. See also Daniel P. Marshall, *Those Who Fell from the Sky: A History of the Cowichan People* (Duncan, BC: Cowichan Tribes, 1999).

10 McGregor Young, "Memo as to restrictions upon fishing in B.C.," 28 July 1913, in "Memorandum of Material of Record with the Royal Commission on Indian Affairs for the Province of British Columbia – Re – Fisheries, Rights, Privileges and Problems of Indians in British Columbia, 1915," *Minutes and Report of the Royal Commission on Indian Affairs for the Province of British Columbia* (call number E78.B9 B754 c. 1), p. 278, Department of Indian Affairs and Northern Development (DIAND) Library.

11 Wetmore, "Regulations re Fishing and Hunting," 6 August 1913, DIA, RG 10, vol. 11026, file FR1 (reel T-3964), LAC.

12 Testimony of Oweekeno spokesperson Joseph Chamberlain, 16 August 1913, DIA, RG 10, vol. 11024, file AH2 (reel T-3961, T-3962), LAC.

13 Testimony of Chief Moody Humchit with questions from Commission Chair E.L. Wetmore, 25 August 1913, ibid.

14 See the discussion in Chapter 7.

15 Testimony of Chief Pierre Michel, Okanagan Agency Evidence, DIA, RG 10, vol. 11025, file AH9 (reel T-3963), p. 28, LAC.

16 Minutes of a Conference between the Royal Commission, the Dominion and Provincial Fisheries Departments, and the Department of Indian Affairs, 9 April 1914, in Royal Commission on Indian Affairs for the Province of British Columbia, Evidence, vol. 1, pp. 413-27, archived at UBC Library, Special Collections. Attendees: Commissioners Macdowall, Shaw, and McKenna; Dominion Chief Inspector of Fisheries F.H. Cunningham, and Inspectors John C.T. Williams and E.G. Taylor; Provincial Deputy Commissioner of Fisheries D.N. McIntyre and Assistant Commissioner J.P. Babcock; and Department of Indian Affairs Inspector W.E. Ditchburn and Indian Agent W.R. Robertson.

17 Ibid., p. 423.

18 Ibid., p. 424.

19 Testimony of Chief Dan Watts, Hupacasath (Opetchisaht) Tribe, West Coast Agency Evidence, 11 May 1914, DIA, RG 10, vol. 11025, file AH13 (reel T-3964), pp. 37-38, LAC.

20 Testimony of Tatoosh, Hupacasath (Opetchisaht) Tribe, 11 May 1914, ibid., pp. 45-46.

21 Testimony of Tseshaht spokesperson Mr. Bill, 11 May 1914, ibid., pp. 52-53.
22 Testimony of Chief Charlie Jackson, Uchuklesaht Tribe, 13 May 1914, ibid., pp. 61-62.
23 Transcript of Royal Commission Meeting with Representatives of the Dominion and Provincial Fisheries Officials in Regard to Fishing Privileges of Indians in BC, 23 December 1915, DIA, RG 10, vol. 3908, file 107297-2 (reel C-10160), LAC. Attendees: Commission Chair White and Commissioners Carmichael, Macdowall, McKenna, Shaw; Dominion Chief Inspector of Fisheries Cunningham and Inspectors J.T. Williams (District 2), W.M. Halliday (District 1), and E.G. Taylor (District 3); and Provincial Assistant Commissioner of Fisheries J. P. Babcock and Deputy Commissioner D.N. McIntyre.
24 "Report Re Material on Record with the Royal Commission Concerning Fisheries Rights, Privileges and Problems of Indians in British Columbia," DIA, RG 10, vol. 11026, file FR1 (reel T-3964), LAC.
25 Transcript of Royal Commission Meeting with Representatives of the Dominion and Provincial Fisheries Officials, pp. 5-6. See note 24 for list of attendees.
26 Ibid., pp. 6-7.
27 N.W. White, "Memorandum Re Fishing Rights and Privileges of Indians in B.C.," 12 January 1916, in *Minutes and Report of the Royal Commission on Indian Affairs for the Province of British Columbia*, DIAND Library.
28 The Union of British Columbia Indian Chiefs has made the full text of the report available online at http://www.ubcic.bc.ca/Resources/final_report.htm (accessed 20 June 2007).
29 Cole Harris, "How Did Colonialism Dispossess? Comments from an Edge of Empire," *Annals of the Association of American Geographers* 94 (2004): 176.
30 C. Harris, *Making Native Space,* 247.
31 Royal Commission on Indian Affairs, New Westminster Agency Evidence, 15 February 1916, DIA, RG 10, series A, vol. 1450 (reel C-14264), pp. 638-45, LAC.
32 Scott to Macdowall, 2 May 1916, DIA, RG 10, vol. 3822, file 59335-1 (reel C-10144), LAC.
33 Macdowall to Scott, 9 May 1916, ibid.
34 Royal Commission on Indian Affairs for the Province of British Columbia, *Report of the Royal Commission on Indian Affairs for the Province of British Columbia* (Victoria: Acme Press, 1916). For the Bella Coola Agency, this resolution appeared on p. 272; Kamloops Agency, p. 356; Kwawkewlth Agency, p. 440; Lytton Agency, pp. 542-43; Nass Agency, pp. 621-22; New Westminster Agency, p. 694; Queen Charlotte Agency p. 742; West Coast Agency, p. 907; Williams Lake Agency, p. 956.
35 Royal Commission on Indian Affairs for the Province of British Columbia, *Confidential Report of the Royal Commission on Indian Affairs for the Province of British Columbia* (Victoria: Acme Press, 1916).
36 Ibid., p. 18.
37 Ibid., p. 21.
38 Ibid., p. 22.
39 See Chapter 6.
40 Statement of the Committee of the Allied Tribes of British Columbia for the Government of Canada, 5 February 1919, DIA, RG 10, vol. 7782, file 27150-3-5B; Statement of the United Tribes of Northern British Columbia for the Government of Canada, 27 February 1919, DIA, RG 10, vol. 7784, file 27150-3-13, pt. 1 reel C-12064); and Statement of the Allied Indian Tribes of British Columbia for the Government of Canada, 12 November 1919, DIA, RG 10, vol. 3821, file 59,335, pt. 4A, LAC.
41 *Indian Affairs Settlement Act,* S.B.C. 1919, c. 32.
42 *Indian Act,* R.S.C. 1906, c. 81, s. 49.
43 *British Columbia Indian Lands Settlement Act,* S.C. 1920, c. 51, s. 3.

44 Scott to Ditchburn, and Scott to Teit, 6 October 1920, DIA, RG 10, C-II-2, vol. 11302 (reel T-16114), LAC.
45 Ditchburn to Scott, 28 November 1922, DIA, RG 10, vol. 7784, file 27150-3-13, pt. 1 (reel C-12064), LAC.
46 Ditchburn to Scott, 17 January 1923, ibid.
47 Clark to T.D. Pattullo, minister of Lands, 17 March 1923, ibid.
48 C. Harris, *Making Native Space,* 254.
49 British Columbia Order-in-Council 911/23, 26 July 1923.
50 Transcript of Conference between the Department of Indian Affairs and the Executive Committee of the Allied Indian Tribes of British Columbia, 7-8 August 1923, RG 10, vol. 3821, file 59335, part 4A (reel C-10143), LAC. Attendees: Deputy Superintendent General of Indian Affairs Duncan C. Scott and Chief Inspector of Indian Affairs for British Columbia W.E. Ditchburn; Allied Indian Tribes of British Columbia chairman, Rev. P.R. Kelly (Haida), and secretary, Andrew Paull (Squamish); Ambrose Reid of the United Tribes of Northern BC; Alec Leonard (Kamloops), Thos. Adolph (Fountain), Narcisse Batiste (Okanagan), and Stephen Retasket (Lillooet) representing the Interior Tribes of British Columbia; Geo. Matheson and Simon Pierre of the Lower Fraser Tribes; Chris Paul (Saanich); John Elliot (Cowichan); Mrs. Cook (Kwawkewlth); and A.E. O'Meara, general counsel of the Allied Tribes.
51 Ibid., pp. 135-66.
52 *Alaska Pacific Fisheries v. United States* (1918), 248 U.S. 78.
53 Transcript of Conference between the Department of Indian Affairs and the Executive Committee of the Allied Indian Tribes, pp. 166-70.
54 Ibid., p. 198.
55 Canada Order-in-Council, PC 1265-24, 19 July 1924.
56 Hamar Foster, "Letting Go the Bone: The Idea of Indian Title in British Columbia, 1849-1927," in *Essays in the History of Canadian Law,* vol.6, *British Columbia and the Yukon,* ed. Hamar Foster and John McLaren (Toronto: Osgoode Society for Canadian Legal History, 1995), 28-86.
57 *Indian Act,* R.S.C. 1927, c. 98, s. 141.
58 *Calder v. A.G. B.C.,* [1973] S.C.R. 313, (1973) 34 D.L.R. (3d) 145.
59 *R. v. Sparrow,* [1990] 1 S.C.R. 1075, [1990] 3 C.N.L.R. 160.
60 British Columbia Order-in-Council 1086, 29 July 1938. See Hamar Foster, "Roadblocks and Legal History, Part 1: Do Forgotten Cases Make Good Law?" *The Advocate* 54 (1996): 355.
61 C. Harris, *Making Native Space,* 261.
62 These reserves are mapped in the Appendix and listed in a table in UBC's Information Repository, at http://hdl.handle.net/2429/648.

Conclusion

1 *British Columbia Terms of Union* (May 16, 1871), R.S.C. 1985, app. 2, no. 10, term 13.
2 Cole Harris, *Making Native Space: Colonialism, Resistance, and Reserves in British Columbia* (Vancouver: UBC Press, 2002).
3 Stuart Banner, *How the Indians Lost Their Land: Law and Power on the Frontier* (Cambridge, MA: Harvard University Press, 2005), 236, 240.
4 Sarah Carter, *Lost Harvests: Prairie Indian Reserve Farmers and Government Policy* (Montreal and Kingston: McGill-Queen's University Press, 1990), 149-56.
5 The international and domestic legal regimes governing fisheries are reviewed in the Introduction.
6 Irene Spry, "The Tragedy of the Loss of the Commons in Western Canada," in *As Long as the Sun Shines and Water Flows: A Reader in Canadian Native Studies,* ed. Ian A.L.

Getty and Antoine S. Lussier (Vancouver: UBC Press, 1983), 212. A note on terminology: Spry is using "tragedy of the commons" in this passage to refer to a tragedy of a regime of open access. Otherwise, she uses "common property resources" and "commons" interchangeably to refer to resources and land governed by the traditions and customs of local communities. In this book, I have used the term "common property" differently, to refer to a regime in which individuals have a right not to be excluded.

7 See Hamar Foster, "Letting Go the Bone: The Idea of Indian Title in British Columbia, 1849-1927," in *Essays in the History of Canadian Law,* vol.6, *British Columbia and the Yukon,* ed. Hamar Foster and John McLaren (Toronto: Osgoode Society for Canadian Legal History, 1995), 28-86. For a general statement of that law see Kent McNeil, *Common Law Aboriginal Title* (Oxford: Clarendon Press, 1989).

8 John Comaroff, "Colonialism, Culture, and the Law: An Introduction," *Law and Social Inquiry* 26 (2001): 311.

9 *R. v. Lewis* (20 October 1988), Squamish (B.C. Prov. Ct.) (Testimony of Albert James Ionson, 27 October 1987, Trial Transcript at 126).

10 Ibid., Exhibit 6.

11 Ibid., Exhibit 9, Squamish Indian Band, By-law No. 10, *A By-law for the Preservation, Protection and Management of Fish on the Reserve* (12 September 1977).

12 *R. v. Lewis,* [1989] 4 C.N.L.R. 133 (B.C. Co. Ct.), at para. 42.

13 *R. v. Lewis,* [1996] 1 S.C.R. 921, [1996] 3 C.N.L.R. 131, affirming (1993), 80 B.C.L.R. (2d) 224, [1993] 4 C.N.L.R. 98 (B.C.C.A.), reversing [1989] 4 C.N.L.R. 133 (B.C. Co. Ct.).

14 *R. v. Nikal,* [1996] 1 S.C.R. 1013, [1996] 3 C.N.L.R. 178.

15 *Lewis,* para. 33.

16 *Lewis,* para. 48.

17 *Nikal,* para. 83.

18 Royal Commission on Indian Affairs for the Province of British Columbia, *Report of the Royal Commission on Indian Affairs for the Province of British Columbia* (Victoria: Acme Press, 1916), p. 637.

19 See J. Michael Thoms, "Ojibwa Fishing Grounds: A History of Ontario Fisheries Law, Science, and the Sportsmen's Challenge to Aboriginal Treaty Rights, 1650-1900" (PhD diss., Department of History, University of British Columbia, 2004); Peggy J. Blair, 'Settling the Fisheries: Pre-Confederation Crown Policy in Upper Canada and the Supreme Court's Decisions in *R. v. Nikal* and *Lewis,*" *Revue générale de droit* 31 (2001): 87-172. For a different view, and the source of much of the Supreme Court of Canada's evidence, see Roland Wright, "The Public Right of Fishing, Government Fishing Policy, and Indian Fishing Rights in Upper Canada," *Ontario History* 86 (1994): 337. See also Chapter 4.

20 This clause, the single most important indicator of pre-Confederation fisheries policy in British Columbia, and the legal touchstone for those who sought protection for Native fishing rights, is not mentioned in either *Nikal* or *Lewis.* It is a serious omission.

21 See Chapter 8.

22 *Alaska Pacific Fisheries v. United States* (1918), 248 U.S. 78, 39 S. Ct. 40 at 41, 4 Alaska Fed. 709.

23 Ibid., at 42.

24 *R. v. Jack,* (1979), [1980] 1 S.C.R. 294, [1979] 2 C.N.L.R. 195, at para. 30.

25 *Lewis,* para. 53.

26 There is a large body of related case law in the United States flowing from *Winters v. United States* (1908), 207 U.S. 564, 28 S.Ct. 207, 52 L.Ed. 340. In this case involving access to water, the US Supreme Court determined that the grant of a reservation included an implied grant of water rights because without water rights the reservation was

not viable. Known as "the *Winters* doctrine," it was a more general enunciation of the principle underlying the decision in *Alaska Pacific Fisheries:* Indian reservations included, even by implication, the resources necessary to make viable the purposes for which the reservation had been allotted.

27 *R. v. Marshall,* [1999] 3. S.C.R. 456, [1999] 4 C.N.L.R. 161. The Supreme Court followed the original decisions with additional reasons for judgment when it denied an application for a rehearing and stay of the decision in *R. v. Marshall,* [1999] 3 S.C.R. 533, [1999] 4 C.N.L.R. 301. See Ken Coates, *The Marshall Decision and Native Rights* (Montreal and Kingston: McGill-Queen's University Press, 2000); William C. Wicken, *Mi'kmaq Treaties on Trial: History, Land, and Donald Marshall Junior* (Toronto: University of Toronto Press, 2002).

28 *R. v. Huovinen* (2000), 188 D.L.R. (4th) 28, (2000) 140 B.C.A.C. 260, leave to appeal refused [2000] S.C.C.A. No. 478 (Supreme Court of Canada, 29 March 2001).

29 *R. v. Kapp,* [2006] 10 W.W.R. 577, [2006] 3 C.N.L.R. 282, leave to appeal allowed [2006] S.C.C.A. No. 331 (Supreme Court of Canada, 14 December 2006).

30 See *Nisga'a Final Agreement,* Canada, British Columbia, and Nisga'a Nation, 27 April 1999, Chapter 8; *Tsawwassen Final Agreement,* Canada, British Columbia, and Tsawwassen Nation, 8 December 2006, Chapter 7.

Bibliography

Archival Sources

British Columbia Archives (Victoria)
British Columbia Attorney General, GR-0429
British Columbia Department of Fisheries, GR-0435
Great Britain, Colonial Office Correspondence, GR-1486
British Columbia County Court (Vancouver), GR-1845
Scottish Records Office, MS-1640

Department of Indian Affairs and Northern Development (Vancouver)
Indian Reserve Commission. Federal Collection of Minutes of Decision, Correspondence and Sketches.
Royal Commission on Indian Affairs for the Province of British Columbia, 1913-16.
  Minutes and Report of the Royal Commission on Indian Affairs for the Province of British Columbia (call number E78.B9 B754 c. 1).

Library and Archives Canada (Ottawa)
Department of Indian Affairs, Record Group 10
Department of Marine and Fisheries, Record Group 23
Records of the Privy Council Office, Record Group 2

Pacific Region Federal Records Centre (Burnaby)
Department of Marine and Fisheries, Record Group 23

University of British Columbia Library, Rare Books and Special Collections (Vancouver)
Contract between Paul Sebassah and C.S. Windsor, 23 June 1877.
Contract between Paul Sebassah and the North West Commercial Company, 14 June 1878.
Dominion-British Columbia Fisheries Commission, 1905-7, Evidence.
Ford, Lillian, and Cole Harris. "B.C. Colonial Indian Reserves." Unpublished paper, April 1999, archived at UBC Special Collections.
Kennedy, Dorothy I.D., and Randy Bouchard. *Utilization of Fish, Beach Foods, and Marine Mammals by the Squamish Indian People of British Columbia*. Victoria: British Columbia Indian Language Project, 1976.
Royal Commission on Indian Affairs for the Province of British Columbia, 1913-16, Evidence.

Cases

A.G. B.C. v. A.G. Canada, [1914] A.C. 153, (1913) 15 D.L.R. 308 (H.L.).
A.G. Canada v. A.G. Ontario, [1898] A.C. 700, (1898) 12 C.R.A.C. 48 (H.L.).
Alaska Pacific Fisheries v. United States (1918), 248 U.S. 78.
Calder v. A.G. B.C., [1973] S.C.R. 313, (1973) 34 D.L.R. (3d) 145.
Canada v. Robertson (1882), 6 S.C.R. 52.
Delgamuukw v. British Columbia, [1997] 3 S.C.R. 1010, [1998] 1 C.N.L.R. 14.
Dixon v. Snetsinger (1873), 23 U.C.C.P. 235.
Haida Nation v. British Columbia (Minister of Forests), 2004 SCC 73, [2004] 3 S.C.R. 511.
In Re Provincial Fisheries, [1895] 26 S.C.R. 444.
Malcolmson v. O'Dea (1863), 10 H.L. Cas. 593.
North v. Kendall (14 December 1904) (B.C. Supreme Court), British Columbia Archives, GR-0429, box 16, file 2.
R. v. Adams, [1996] 3 S.C.R. 101, [1996] 4 C.N.L.R. 1.
R. v. Bob, [1979] 4 C.N.L.R. 71 (B.C.C.C.).
R. v. Côte, [1996] 3 S.C.R. 139, [1996] 4 C.N.L.R. 26.
R. v. Gladstone, [1996] 2 S.C.R. 723, [1996] 4 C.N.L.R. 65.
R. v. Huovinen (2000), 188 D.L.R. (4th) 28, (2000) 140 B.C.A.C. 260, leave to appeal refused [2000] S.C.C.A. No. 478 (Supreme Court of Canada, 29 March 2001).
R v. Jack (1979), [1980] 1 S.C.R. 294, [1979] 2 C.N.L.R. 25.
R. v. Jimmy (1987), B.C.L.R. (2d) 145 (B.C.C.A.).
R. v. Kapp, [2003] 4 C.N.L.R. 238.
R. v. Kapp, [2006] 10 W.W.R. 577, [2006] 3 C.N.L.R. 282, leave to appeal allowed [2006] S.C.C.A. No. 331 (Supreme Court of Canada, 14 December 2006).
R. v. Lewis (20 October 1988), Squamish, (B.C. Prov. Ct.).
R. v. Lewis, [1989] 4 C.N.L.R. 133 (B.C. Co. Ct.).
R. v. Lewis, [1996] 1 S.C.R. 921, [1996] 3 C.N.L.R. 131.
R. v. Marshall, [1999] 3 S.C.R. 456, [1999] 4 C.N.L.R. 161.
R. v. Marshall, [1999] 3 S.C.R. 533, [1999] 4 C.N.L.R. 301 (motion for hearing and stay).
R. v. N.T.C. Smokehouse Ltd., [1996] 2 S.C.R. 672, [1996] 4 C.N.L.R. 130.
R. v. Nikal, [1996] 1 S.C.R. 1013, [1996] 3 C.N.L.R. 178.
R. v. Sparrow, [1990] 1 S.C.R. 1075, [1990] 3 C.N.L.R. 160.
R. v. Van der Peet, [1996] 2 S.C.R. 507, [1996] 4 C.N.L.R. 177.
Rex v. Charley (14 December 1915) (B.C. County Court, Vancouver), RG 10, vol. 10899, file 167/20-2, Library and Archives Canada.
Taku River Tlingit v. British Columbia (Project Assessment Director), 2004 SCC 74, [2004] 3 S.C.R. 550.
Tsilhqot'in Nation v. British Columbia, 2007 B.C.S.C. 1700.
United States v. State of Washington 384 F. Supp. 312.
Wewaykum Indian Band v. Canada, [2002] 4 S.C.R. 245; [2003] 1 C.N.L.R. 341.
Winters v. United States (1908), 207 U.S. 564, 28 S.Ct. 207, 52 L.Ed. 340.

Articles, Books, and Other Documents

Adachi, Ken. *The Enemy That Never Was: A History of the Japanese Canadians.* Toronto: McClelland and Stewart, 1991.
Alcantara, Christopher. "Individual Property Rights on Canadian Indian Reserves: The Historical Emergence and Jurisprudence of Certificates of Possession." *Canadian Journal of Native Studies* 23 (2003): 391-424.

Anderson, A.C. "Notes on the Indian Tribes of British North America, and the Northwest Coast." *Historical Magazine* 7 (March 1863): 73-81.

Anderson, Kay. *Vancouver's Chinatown: Racial Discourse in Canada, 1875-1980.* Montreal and Kingston: McGill-Queen's University Press, 1991.

Arnett, Chris. *The Terror of the Coast: Land Alienation and Colonial War on Vancouver Island and the Gulf Islands, 1849-1863.* Burnaby, BC: Talonbooks, 1999.

Banner, Stuart. *How the Indians Lost Their Land: Law and Power on the Frontier.* Cambridge, MA: Harvard University Press, 2005.

Barman, Jean. *The West beyond the West: A History of British Columbia.* Rev. ed. Toronto: University of Toronto Press, 1996.

Blackstone, William. *Commentaries on the Laws of England.* A facsimile of the first edition of 1765-69. Chicago: University of Chicago Press, 1979.

–. *Commentaries on the Laws of England.* Edited by Wayne Morrison. London: Cavendish, 2001.

Blair, Peggy J. "'For Our Race Is Our Licence': Culture, Courts and Conflict over Aboriginal Hunting and Fishing Rights in Southern Ontario." LLD diss., University of Ottawa, 2003.

–. "Settling the Fisheries: Pre-Confederation Crown Policy in Upper Canada and the Supreme Court's Decisions in *R. v. Nikal* and *Lewis.*" *Revue générale de droit* 31 (2001): 87-172.

–. "Solemn Promises and *Solum* Rights: The Saugeen Ojibway Fishing Grounds and *R. v. Jones* and *Nadjiwon.*" *Ottawa Law Review* 28 (1996-97): 125-43.

Blomley, Nicholas K. "Law, Property, and the Geography of Violence: The Frontier, the Survey, and the Grid." *Annals of the Association of American Geographers* 93 (2003): 121-41.

–. *Law, Space, and the Geographies of Power.* New York: The Guilford Press, 1994.

Bogue, Margaret Beattie. *Fishing the Great Lakes: An Environmental History, 1783-1933.* Madison: University of Wisconsin Press, 2000.

Bowsfield, Hartwell, ed. *Fort Victoria Letters: 1846-1852.* Winnipeg: Hudson's Bay Record Society, 1979.

Boxberger, Daniel. "In and Out of the Labour Force: The Lummi Indians and the Development of the Commercial Salmon Fishery in North Puget Sound, 1880-1900." *Ethnohistory* 35, 2 (1988): 161-90.

–. *To Fish in Common: The Ethnohistory of Lummi Indian Salmon Fishing.* Seattle: University of Washington Press, 2000. First published in 1989 by University of Nebraska Press.

Brealey, Kenneth. "First (National) Space: (Ab)original (Re)mappings of British Columbia." PhD diss., Department of Geography, University of British Columbia, 2002.

–. "Travels from Point Ellice: Peter O'Reilly and the Indian Reserve System in British Columbia." *BC Studies* 115-16 (1997-98): 181-236.

Brilmayer, Lea, and Natalie Klein. "Land and Sea: Two Sovereignty Regimes in Search of a Common Denominator." *New York University Journal of International Law* 33 (2000-1): 703-68.

Brown, Dennis. *Salmon Wars: The Battle for the West Coast Salmon Fishery.* Madeira Park, BC: Harbour Publishing, 2005.

Bushnell, S.I. "The Use of American Cases." *University of New Brunswick Law Journal* 35 (1986): 157-81.

Cail, Robert E. *Land, Man, and the Law: The Disposal of Crown Lands in British Columbia, 1871-1913.* Vancouver: UBC Press, 1974.

Carlson, Keith Thor. "History Wars: Considering Contemporary Fishing Site Disputes." In *A Stó:lō Coast Salish Historical Atlas,* edited by Keith Thor Carlson, 58-59. Vancouver: Douglas and McIntyre, 2001.

–. "Innovation, Tradition, Colonialism and Aboriginal Fishing Conflicts in the Lower Fraser Canyon." In *New Histories for Old: Changing Perspectives on Canada's Native Pasts*, edited by Ted Binnema and Susan Neylan, 145-74. Vancouver: UBC Press, 2007.

–. "Rethinking Dialogue and History: The King's Promise and the 1906 Aboriginal Delegation to London." *Native Studies Review* 16, 2 (2005): 1-38.

Carter, Sarah. *Lost Harvests: Prairie Indian Reserve Farmers and Government Policy.* Montreal and Kingston: McGill-Queen's University Press, 1990.

Clayton, Daniel. *Islands of Truth: The Imperial Fashioning of Vancouver Island.* Vancouver: UBC Press, 2000.

Coates, Ken. *The Marshall Decision and Native Rights.* Montreal and Kingston: McGill-Queen's University Press, 2000.

Cohen, Felix. "Dialogue on Private Property." *Rutgers Law Review* 9 (1954): 357-87.

Cohen, Morris. "Property and Sovereignty." *Cornell Law Quarterly* 13 (1927-28): 8-30.

Comaroff, John. "Colonialism, Culture, and the Law: An Introduction." *Law and Social Inquiry* 26 (2001): 305-14.

Côté, J.E. "The Reception of English Law." *Alberta Law Review* 15 (1977): 29-92.

Cronin, William. *Changes in the Land: Indians, Colonists, and the Ecology of New England.* New York: Hill and Wang, 1983.

Dicey, A.V. "The Paradox of the Land Law." *Law Quarterly Review* 21 (1905): 221-32.

Dudas, Jeffrey R. "Law at the American Frontier." *Law and Social Inquiry* 29 (2004): 859-90.

Duff, Wilson. "The Fort Victoria Treaties." *BC Studies* 3 (1969): 3-57.

–. *The Impact of the White Man.* 3rd ed. Victoria: Royal British Columbia Museum, 1997.

Dunlop, H.F. *Andy Paull: As I Knew Him and Understood His Times.* Vancouver: Standard Press, 1983.

Elliot, Dave, Sr. *Salt Water People.* Saanich, BC: School District No. 63 (Saanich), 1983.

Evenden, Matthew D. *Fish versus Power: An Environmental History of the Fraser River.* Cambridge: Cambridge University Press, 2004.

Fisher, Robin. *Contact and Conflict: Indian-European Relations in British Columbia, 1774-1890.* 2nd ed. Vancouver: UBC Press, 1992.

–. "Joseph Trutch and Indian Land Policy." *BC Studies* 12 (1971-72): 3-33.

Forester, Joseph E., and Anne D. Forester. *Fishing: British Columbia's Commercial Fishing History.* Saanichton, BC: Hancock House, 1975.

Foster, Hamar. "Letting Go the Bone: The Idea of Indian Title in British Columbia, 1849-1927." In *Essays in the History of Canadian Law.* Vol. 6, *British Columbia and the Yukon,* edited by Hamar Foster and John McLaren, 28-86. Toronto: Osgoode Society for Canadian Legal History, 1995.

–. "Roadblocks and Legal History, Part 1: Do Forgotten Cases Make Good Law?" *The Advocate* 54 (1996): 355-66.

–. "A Romance of the Lost: The Role of Tom MacInnes in the History of the British Columbia Indian Land Question." In *Essays in the History of Canadian Law.* Vol. 8, *In Honour of R.C.B. Risk,* edited by G. Blaine Baker and Jim Phillips, 171-212. Toronto: Osgoode Society for Canadian Legal History, 1999.

–. "The Saanichton Bay Marina Case: Imperial Law, Colonial History and Competing Theories of Aboriginal Title." *UBC Law Review* 23 (1989): 629-50.

Foster, Hamar, and Alan Grove. "'Trespassers on the Soil': *United States v. Tom* and a New Perspective on the Short History of Treaty Making in Nineteenth-Century British Columbia." *BC Studies* 138-39 (2003): 51-84.

Galois, Robert. "Colonial Encounters: The Worlds of Arthur Wellington Clah." *BC Studies* 115-16 (1997-98): 105-47.

–. "The Indian Rights Association, Native Protest Activity and the 'Land Question' in British Columbia, 1903-1916." *Native Studies Review* 8 (1992): 113-83.

Grey, T.C. "The Disintegration of Property." In *Property: Nomos 12,* edited by J.R. Penncock and J.W. Chapman, 69-85. New York: New York University Press, 1980.

Grotius, Hugo. *The Freedom of the Seas, or, the Right Which Belongs to the Dutch to Take Part in the East Indian Trade.* Translated by Ralph Magoffin. Edited by James Brown Scott. 1608. New York: Oxford University Press, 1916.

Guha, Ranajit. *A Rule of Property for Bengal: An Essay on the Idea of Permanent Settlement.* Durham, NC: Duke University Press, 1996. First published 1963 by Mouton (Paris).

Hale, Matthew. "De Juris Marie et Brachiorum Ejusdem." In *A Collection of Tracts Relative to the Law of England,* Vol. 1, Part 1, edited by F. Hargrave, 5-44. 1787. Abington, UK: Professional Books, 1982.

Hansen, Lise C. "Treaty Rights and the Development of Fisheries Legislation in Ontario: A Primer." *Native Studies Review* 7 (1991): 1-21.

Harris, Cole. "How Did Colonialism Dispossess? Comments from an Edge of Empire." *Annals of the Association of American Geographers* 94 (2004): 165-82.

–. *Making Native Space: Colonialism, Resistance, and Reserves in British Columbia.* Vancouver: UBC Press, 2002.

–. *The Resettlement of British Columbia: Essays on Colonialism and Geographical Change.* Vancouver: UBC Press, 1997.

Harris, Cole, and Robert Galois. "A Population Geography of British Columbia in 1881." In *The Resettlement of British Columbia: Essays on Colonialism and Geographical Change,* by Cole Harris, 137-60. Vancouver: UBC Press, 1997.

Harris, Douglas C. "The Boldt Decision in Canada: Aboriginal Treaty Rights to Fish on the Pacific." In *The Power of Promises: Perspectives on Treaties with Native Peoples in the Pacific Northwest,* edited by Alexandra Harmon. Seattle: University of Washington Press, 2008.

–. *Fish, Law, and Colonialism: The Legal Capture of Salmon in British Columbia.* Toronto: University of Toronto Press, 2001.

–. "Indian Reserves, Aboriginal Fisheries, and the Public Right to Fish." In *Despotic Dominion: Property Rights in British Settler Colonies,* edited by John McLaren, A.R. Buck, and Nancy E. Wright, 266-93. Vancouver: UBC Press, 2004.

–. "The Nlha7kapmx Meeting at Lytton, 1879, and the Rule of Law." *BC Studies* 108 (1995-96): 5-25.

Harvey, Christopher. Review of *Fish, Law, and Colonialism: The Legal Capture of Salmon in British Columbia* by Douglas C. Harris. *The Advocate* 61 (2003): 125-27.

Hay, Douglas, and Paul Craven, eds. *Masters, Servants, and Magistrates in Britain and the Empire, 1562-1955.* Durham: University of North Carolina Press, 2004.

Hay, Douglas, and Nicholas Rogers. *Eighteenth-Century English Society: Shuttles and Swords.* Oxford: Oxford University Press, 1997.

Hayden, Brian, ed. *A Complex Culture of the British Columbia Plateau: Traditional Stl'átl'imx Resource Use.* Vancouver: UBC Press, 1992.

Higginbottom, Edward N. "The Changing Geography of the Salmon Canning Industry in British Columbia, 1870-1931." Master's thesis, Simon Fraser University, 1988.

House of Assembly Correspondence Book, August 12th 1856 to July 6th 1859. Victoria: W. Cullin, Printer to the King, 1918.

Kennedy, Dorothy I.D., and Randy Bouchard. "*Stl'átl'imx* (Fraser River Lillooet) Fishing." In *A Complex Culture of the British Columbia Plateau: Traditional* Stl'átl'imx *Resource Use,* edited by Brian Hayden, 266-354. Vancouver: UBC Press, 1992.

Kew, Michael. "Salmon Availability, Technology, and Adaptation." In *A Complex Culture of the British Columbia Plateau: Traditional* Stl'átl'imx *Resource Use,* edited by Brian Hayden, 177-221. Vancouver: UBC Press, 1992.

Knight, Rolf. *Indians at Work: An Informal History of Native Indian Labour in British Columbia, 1858-1930.* Vancouver: New Star Books, 1978; rev. ed. 1996.

Laforet, Andrea, and Annie York. *Spuzzum: Fraser Canyon Histories, 1808-1939.* Vancouver: UBC Press, 1998.

Lambert, Thomas Wilson. *Fishing in British Columbia.* London: H. Cox, 1907.

Letter from the Methodist Missionary Society to the Superintendent-General of Indian Affairs Respecting British Columbia Troubles. Toronto: Methodist Church of Canada, 1889.

Locke, John. *Two Treatises of Government.* Edited by Peter Laslett. Cambridge: Cambridge University Press, 1960.

Loo, Tina. *Making Law, Order, and Authority in British Columbia, 1821-1871.* Toronto: University of Toronto Press, 1994.

Lutz, John. "Seasonal Rounds in an Industrial World." In *A Stó:lō -Coast Salish Historical Atlas,* edited by Keith Thor Carlson, 64-67. Vancouver: Douglas and McIntyre, 2001.

–. "Work, Wages and Welfare in Aboriginal-Non-Aboriginal Relations, British Columbia, 1849-1970." PhD diss., Department of History, University of Ottawa, 1994.

Lyons, Cicely. *Salmon: Our Heritage.* Vancouver: British Columbia Packers, 1969.

Lytwyn, Victor P. "Ojibway and Ottawa Fisheries around Manitoulin Island: Historical and Geographical Perspectives on Aboriginal and Treaty Rights." *Native Studies Review* 6 (1990): 1-30.

–. "The Usurpation of Aboriginal Fishing Rights: A Study of the Saugeen Nation's Fishing Islands in Lake Huron." In *Co-Existence? Studies in Ontario-First Nations Relations,* edited by B.W. Hodgins, S. Heard, and J.S. Milloy, 84-103. Peterborough, ON: Frost Centre for Canadian Heritage and Development Studies, 1992.

Mackie, Richard. "The Colonization of Vancouver Island, 1849-1858." *BC Studies* 96 (1992-93): 3-40.

–. *Trading beyond the Mountains: The British Fur Trade on the Pacific, 1763-1843.* Vancouver: UBC Press, 1996.

MacPherson, C.B. "Capitalism and the Changing Conception of Property." In *Feudalism, Capitalism and Beyond,* edited by Eugene Kamenka and R.S. Neale, 105-24. Canberra: ANU Press, 1975.

–. *Property: Mainstream and Critical Positions.* Toronto: University of Toronto Press, 1978.

Manuel, George, and Michael Posluns. *The Fourth World: An Indian Reality.* Don Mills, ON: Collier-Macmillan, 1974.

Marlatt, Daphne, ed. *Steveston Recollected: A Japanese-Canadian History.* Victoria: Provincial Archives of British Columbia, 1975.

Marshall, Daniel P. "Claiming the Land: Indians, Goldseekers, and the Rush to British Columbia." PhD diss., Department of History, University of British Columbia, 2000.

–. *Those Who Fell from the Sky: A History of the Cowichan People.* Duncan, BC: Cowichan Tribes, 1999.

McHugh, P.G. *Aboriginal Societies and the Common Law: A History of Sovereignty, Status, and Self-Determination.* Oxford: Oxford University Press, 2004.

McNeil, Kent. *Common Law Aboriginal Title.* Oxford: Clarendon Press, 1989.

Meggs, Geoff. *Salmon: The Decline of the British Columbia Fishery.* Vancouver: Douglas and McIntyre, 1995.

Merrill, Thomas W. "Property and the Right to Exclude." *Nebraska Law Journal* 77 (1998): 730-55.

Merry, Sally Engle. "Law and Colonialism." *Law and Society Review* 25 (1991): 889-922.

Miller, Bruce G., ed. *Be of Good Mind: Essays on the Coast Salish*. Vancouver: UBC Press, 2007.

Morton, James W. *Capilano: The Story of a River*. Toronto: McClelland and Stewart, 1970.

Munsche, P.B. *Gentlemen and Poachers: The English Game Laws 1671-1831*. Cambridge: Cambridge University Press, 1981.

Muszynski, Alicja. *Cheap Wage Labour: Race and Gender in the Fisheries of British Columbia*. Montreal and Kingston: McGill-Queen's University Press, 1996.

Newell, Dianne. "Dispersal and Concentration: The Slowly Changing Spatial Pattern of the British Columbia Salmon Canning Industry." *Journal of Historical Geography* 14 (1988): 22-36.

–. *Tangled Webs of History: Indians and the Law in Canada's Pacific Coast Fisheries*. Toronto: University of Toronto Press, 1993.

Papers Connected with the Indian Land Question, 1850-1875, 1877. A facsimile of a document printed in 1875, with a supplement published in 1877. Victoria: Queen's Printers, 1987.

Parenteau, Bill. "'Care, Control and Supervision': Native People in the Canadian Atlantic Salmon Fishery." *Canadian Historical Review* 79 (1998): 1-35.

Patterson, E. Palmer, II. "Andrew Paull and the Canadian Indian Resurgence." PhD diss., University of Washington, 1962.

Pinkerton, E. "Indians in the Fishing Industry." In *Uncommon Property: The Fishing and Fish Processing Industries in British Columbia*, edited by P. Marchak, N. Guppy, and J. McCullan, 249-69. Toronto: Methuen, 1987.

Plucknett, Theodore F.T. *A Concise History of the Common Law*. 5th ed. Boston: Little, Brown, 1956.

Prichard, M.F. Floyd, ed. *The Collected Works of Edward Gibbon Wakefield*. Auckland: Collins, 1969.

Pritchard, John Charles. "Economic Development and Disintegration of Traditional Culture among the Haisla." PhD diss., Department of Anthropology and Sociology, University of British Columbia, 1977.

Raibmon, Paige. "Un Making Native Space: A Genealogy of Indian Policy, Settler Practice and the Micro-Techniques of Dispossession." In *The Power of Promises: Perspectives on Treaties with Native Peoples in the Pacific Northwest*, edited by Alexandra Harmon. Seattle: University of Washington Press, fortheoming 2008.

Ralston, Keith. "The 1900 Strike of Fraser River Sockeye Salmon Fishermen." Master's thesis, Department of History, University of British Columbia, 1965.

Rathbun, Richard. *A Review of the Fisheries in the Contiguous Waters of the State of Washington and British Columbia*. Washington, 1899.

Ray, Arthur J. "Periodic Shortages, Native Welfare, and the Hudson's Bay Company 1670-1930." In *The Subarctic Fur Trade: Native Social and Economic Adaptations*, edited by Shepard Krech III, 1-20. Vancouver: UBC Press, 1984.

Reid, David. "The Development of the Fraser River Salmon Canning Industry, 1885 to 1913." Prepared for Department of the Environment, Fisheries and Marine Service, Pacific Region (July 1973).

Roine, Chris. "The Squamish Aboriginal Economy, 1860-1940." Master's thesis, Department of History, Simon Fraser University, 1996.

Romanoff, Steven. "Fraser Lillooet Salmon Fishing." In *A Complex Culture of the British Columbia Plateau: Traditional* Stl'átl'imx *Resource Use*, edited by Brian Hayden, 470-505. Vancouver: UBC Press, 1992.

Rounsefell, George A., and George B. Kelez. *The Salmon and Salmon Fisheries of Swiftsure Bank, Puget Sound, and the Fraser River.* Washington DC: US Government Printing Office, 1938.

Royal Commission on Indian Affairs for the Province of British Columbia. *Confidential Report of the Royal Commission on Indian Affairs for the Province of British Columbia.* Victoria: Acme Press, 1916.

–. *Report of the Royal Commission on Indian Affairs for the Province of British Columbia.* Victoria: Acme Press, 1916.

Ruggie, John Gerard. "Territoriality and Beyond: Problematizing Modernity in International Relations." *International Organization* 47, 1 (1993): 139-74.

Sack, Robert David. *Human Territoriality: Its Theory and History.* Cambridge: Cambridge University Press, 1986.

Said, Edward W. *Culture and Imperialism.* New York: Vintage Books, 1994.

Sanders, Deidre, Naneen Stuckey, Kathleen Mahoney, and Leland Donald. "What the People Said: Kwakwaka'wakw, Nuu-chah-nulth, and Tsimshian Testimonies before the Royal Commission on Indian Affairs for the Province of British Columbia (1913-1916)." *Canadian Journal of Native Studies* 19 (1999): 213-48.

Sandwell, R.W. *Contesting Rural Space: Land Policy and Practices of Resettlement on Saltspring Island, 1859-1891.* Montreal and Kingston: McGill-Queen's University Press, 2005.

Scott, James C. *Seeing like a State: How Certain Schemes to Improve the Human Condition Have Failed.* New Haven, CT: Yale University Press, 1998.

Seed, Patricia. *Ceremonies of Possession in Europe's Conquest of the New World, 1492-1640.* Cambridge: Cambridge University Press, 1995.

Seymour, Anne Elisabeth. "Natives and Reserve Establishment in Nineteenth-Century British Columbia." Master's thesis, Department of History, University of British Columbia, 1995.

Shankel, G.E. "The Development of Indian Policy in British Columbia." PhD diss., University of Washington, 1945.

Sinclair, Bertrand W. *Poor Man's Rock.* Boston: Little, Brown, 1920.

Singer, Joseph. *Entitlement: The Paradoxes of Property.* New Haven, CT: Yale University Press, 2000.

–. "Sovereignty and Property." *Northwestern University Law Review* 86, 1 (1991-92): 1-56.

Sparrow, Leonna. "Work Histories of a Coast Salish Couple." Master's thesis, University of British Columbia, 1976.

Special Fisheries Commission. *Report of Special Fisheries Commission, 1917.* Ottawa, 1918.

Sproat, Gilbert Malcolm. *The Nootka: Scenes and Studies of Savage Life.* Edited and annotated by Charles Lillard. 1868. Victoria: Sono Nis Press, 1987.

Spry, Irene. "The Tragedy of the Loss of the Commons in Western Canada." In *As Long as the Sun Shines and Water Flows: A Reader in Canadian Native Studies,* edited by Ian A.L. Getty and Antoine S. Lussier, 203-28. Vancouver: UBC Press, 1983.

Starr, Vina A. "Indian Title to Foreshore on Coastal Reserves in British Columbia." Prepared for Department of Indian and Northern Affairs Canada, 28 March 1985.

Stewart, Hilary. *Indian Fishing: Early Methods on the Northwest Coast.* Vancouver: J.J. Douglas, 1977.

Suttles, Wayne. *Coast Salish Essays.* Vancouver: Talonbooks, 1987.

Taylor, Joseph E., III. *Making Salmon: An Environmental History of the Northwest Fisheries Crisis.* Seattle: University of Washington Press, 1999.

Tennant, Paul. *Aboriginal People and Politics: The Indian Land Question in British Columbia, 1849-1989.* Vancouver: UBC Press, 1990.

Thompson, E.P. *Customs in Common: Studies in Traditional Popular Culture*. New York: The New Press, 1993.

–. *Whigs and Hunters: The Origin of the Black Act*. New York: Pantheon Books, 1975.

Thoms, J. Michael. "Ojibwa Fishing Grounds: A History of Ontario Fisheries Law, Science, and the Sportsmen's Challenge to Aboriginal Treaty Rights, 1650-1900." PhD diss., Department of History, University of British Columbia, 2004.

–. "A Place Called Pennask: Fly-Fishing and Colonialism at a British Columbia Lake." *BC Studies* 133 (2002): 69-98.

Titley, Brian E. *A Narrow Vision: Duncan Campbell Scott and the Administration of Indian Affairs in Canada*. Vancouver: UBC Press, 1986.

van Bynkershoek, Cornelius. *De Dominio Maris Dissertatio – The Dissertation on the Sovereignty of the Sea*. Translated by Ralph Magoffin. Edited by James Brown Scott. 1744. New York: Oxford University Press, 1916.

Wadewitz, Lissa. "The Nature of Borders: Salmon and Boundaries in the Puget Sound/Georgia Basin." PhD diss., Department of History, University of California Los Angeles, 2004.

Waldron, Jeremy. *The Right to Private Property*. Oxford: Clarendon Press, 1988.

Walters, Mark. "Aboriginal Rights, *Magna Carta* and Exclusive Rights to Fisheries in the Waters of Upper Canada." *Queen's Law Journal* 23 (1988): 301-68.

Ward, W. Peter. *White Canada Forever: Popular Attitudes and Public Policy towards Orientals in British Columbia*. 2nd ed. Montreal and Kingston: McGill-Queen's University Press, 1990.

Ware, Reuben M. *Five Issues, Five Battlegrounds: An Introduction to the History of Indian Fishing in British Columbia*. Chilliwack, BC: Coqualeetza Education Training Centre, 1983.

Weaver, John. *The Great Land Rush and the Making of the Modern World, 1650-1900*. Montreal and Kingston: McGill-Queen's University Press, 2003.

Wells, Oliver N., ed. *Squamish Legends by Chief August Jack Khahtsahlano and Domanic Charlie*. Vancouver: Charles Chamberlain and Frank T. Coan, 1966.

Wicken, William C. *Mi'kmaq Treaties on Trial: History, Land, and Donald Marshall Junior*. Toronto: University of Toronto Press, 2002.

Wright, Roland. "The Public Right of Fishing, Government Fishing Policy, and Indian Fishing Rights in Upper Canada." *Ontario History* 86 (1994): 337-62.

Yonemura, Hozumi. "Japanese Fishermen in British Columbia and British Fair Play." *Canadian Forum* 10 (1930): 357.

Young, Robert J.C. *Colonial Desire: Hybridity in Theory, Culture and Race*. London: Routledge, 1995.

Ziff, Bruce. "Warm Reception in a Cold Climate: English Property Law and the Suppression of the Canadian Legal Identity." In *Despotic Dominion: Property Rights in British Settler Societies*, edited by John McLaren, A.R. Buck, and Nancy E. Wright, 103-19. Vancouver: UBC Press, 2005.

Index

Note: (t) after a number refers to a table; (f) to a figure or a photograph.

Native policy under Terms of Union, 33, 102, 123, 187; and quantity of land for reserves, 187
Cook's Ferry band, 54
Cory, Justice, 194
Court of Appeal (BC), 193
Cowichan Bay, 148, 149, 150, 152
Cowichan Indian Reserve No. 1, 40-41
Cowichan people, 28; fishing weirs, 168-70; McKenna-McBride commission and, 167-70; and *R. v. Jack,* 196-97; reserve, 152, 167-70; seasonal fishing near Fraser River, 134; and seine fishing, 149, 150
Cowichan River, 134; canneries and, 150; log running on, 26, 40-41; reserves along, 40-41; seine fishers and, 148; sport fishing on, 150; weir fishery on, 26, 41, 168, 169-70, 171
Cowichan Tribes, 148
Cox, Mr. (Indian agent), 173
Crosby, T., 73
Crown lands: Native peoples' rights to use, 25-26; Native purchase/pre-emption of, 12, 155
Cunningham, F.H., 114, 115, 156, 171, 176, 177
Curtis Inlet, 70
customary law: common law vs, 10; European law vs, 9

Dakelh people, 178
Davis, Jack, 128
Deasy, Thomas, 104-5
Dewdney, Edgar, 100
Dickson, Brian, 196-97, 216n55
dip nets, 38, 45, 54, 108(f), 111
Discovery Island, 28, 29(f)
dispossession, 4; common property and, 189-90; extension of sovereignty and, 8; institutional history of, 34; legal regime and, 8, 34; private property rights and, 12; in settler colonies, 9; state law and, 79
Ditchburn, W.E., 181-83, 184
Ditchburn Clark Agreement, 211n17
Dominion-British Columbia Fisheries Commission of 1905-7, 111
Dominion government: bureaucracy, 31; Confederation, and transfer of jurisdic-

tion to, 33-34, 101-2, 102, 187-88; and development of province, 59; fisheries law, 31; and Indians engaged in industry, 37; jurisdiction over fisheries, 33; jurisdiction over Indians, 33; and Native title, 104, 165-66, 185; representatives on JIRC, 37; title to reserves, 185; and treaties, 188
Douglas, James, 21-26, 22(f), 29, 42, 56, 57, 116, 218n25
Douglas, William, 193-94
Douglas band, 68
Douglas Treaties, 5, 20, 21-27, 22(f), 101, 188; fisheries clause, 14, 30, 34, 80, 102, 187, 195; Sproat and, 55, 56
Draney, Robert, 153-54
Draney Fisheries Ltd., 154, 170
Draper, W.H., 32
drift-net fishing: commercial, 109; on Fraser River, 147; limited-licence regime for, 163; regulations governing, 51, 108, 109; seine fishing and, 148, 153; on Squamish River, 192; Wilmot's recommendation regarding, 148
drift-net licences: on central and northern coasts, 138-47; on Fraser River, 131-38; independent, for Native fishers, 184
Duff, Wilson, 24

Edwards, John T., 42
Elhlateese reserve No. 2, 174
Elliot, Dave Sr., 28
Ellis, J.N., 123
enforcement of regulations: during 1870s, 31; and closure of Squamish food fishery, 120; and fishing technologies, 171, 172; Kelly's protest against increase in, 114-15; local circumstances and, 34, 112; and McKenna-McBride commission, 172; against Native fishers, 197; Nlha7kapmx protests against increasing, 57; Nuu-Chah-Nulth on increasing, 172; and selling of fish, 111-12
Esquimalt people, 25
Evenden, Mathew, 114
Exchequer Court of Canada, 86-87
exclusive fisheries: allocation as Crown grants, 90; common law and, 90; Crown recognition of, 84, 194; exclusive rights to land vs, 8; Fisheries and, 85-86, 93,

102; French Crown grants to settlers,
84; Indian Affairs' list of, 101; Indian
reserve commissioners and, 191; and
lakes, 84; land grants and, 89; legality
of, 79; McKenna-McBride commission
on, 180; Miramichi River, 86-88; Mount
Currie, 178; at mouth of Skeena River,
68; on Nass River, 72-75; in non-tidal
waters, 82, 86, 89; O'Reilly's allocations
of, 15, 72-75, 77, 82, 93; ownership
and, 63; parliament and, 32-33; public
right to fish and, 72, 78, 82-85, 86,
89; in public waters, 194; recognition
of vs granting rights to, 79-80; reserve
commissioners' authority to allocate,
85, 98-99; on reserve foreshores, 196;
and rivers, 84; seine-net, 153, 157-60,
163; in tidal waters, 72, 75(f), 81, 90,
97; treaties and, 84; in waters adjacent
to reserves, 194; in waters running
through reserves, 174
exclusive fishing rights: allotments of
reserves and, 187; on Lillooet River,
178; on Nass River, 72; in non-tidal
waters, 80; on reserves, 2, 4, 123-24;
on rivers flowing through reserves, 180,
196; with seine licences on reserve
foreshores, 184
exclusive rights to land: exclusive rights to
fisheries vs, 8; inside fenced property, 19

farming. *See* agriculture
fee simple, 11, 12
fences, 36, 45-46
fish: abundance of, 163, 189; as common
property, 4, 98, 163; competition
for, 28, 101; disappearance of, 197;
inexhaustibility as resource, 20; maxi-
mization to non-Native users, 106;
non-Native interest in, 20; "peddler's
licences" for, 112, 177, 180; private
ownership of, 13; selling of, 111-13,
171, 172-73, 174, 177, 180, 197-98
fisheries: allocation as harvest agreements,
198; allotment of land connected to,
195-96; allotment of reserves and, 30;
"as formerly," 14, 24, 25, 26, 34, 102,
195; capital and, 91, 189; centrality of,
56, 93; in colonial economy, 30; colonial government and, 30-31; as common property, 13, 14, 189, 190; as

common property vs land as private
property, 8; competition for, 31, 35,
110, 127, 132; construction as public
property, 91; Crown ownership in tidal
waters, 81; detachment from reserved
land, 103, 165; division into northern
and southern districts, 130(f); division
of Fraser River fishery from southern
district, 130(f); Douglas Treaties and,
21, 23-27; on Fraser River, 56-57; immigrants and, 25; jurisdiction over, 30;
and land policy, 27-31; location, and
connections with reserves, 28; Native
management of, 6-7; Native peoples'
necessitous claims to vs legal rights, 84;
Native prior claim to, 14; non-Native
interest in, 24, 27-28, 32, 107; as open-
access resource, 127-28; open access to,
4, 127-28; ownership modes, 61, 63;
policies, 34; prior rights to, 80, 189; as
privileges, 101; public, 63; registry of
fishers, 135; as settlements limiting non-
Native settlement, 41-42; and size of
land base, 6, 7; sovereignty over, 12-17;
sustainability of, 132, 197; treaties
and, 191, 198; and viability of reserve
economies, 104; Yale, 30, 56-57. *See
also* exclusive fisheries
Fisheries, Department of (BC), 130-31,
142, 171; McKenna-McBride Commission and, 171, 174, 176-77
Fisheries, Department of (Canada): and
allotment of reserves for access to fisheries, 83-84, 162; and authority to
allocate exclusive fisheries, 85, 98-99;
Barricade Treaty, 103; and cannery
control over licensing, 132-33; and
cannery licences, 132, 134; closure of
Seymour and Lynn Creeks to gaffing,
124-25; confiscation of fishing gear,
125; control of fisheries, 127-28; and
Cowichan fish weirs, 168-70; and
Cowichan fishers on Fraser River, 134;
division of fishery into three districts,
130(f); and drift-net fishing, 148; and
exclusive fisheries, 15, 82, 83, 85-86,
88, 89, 102; and exclusive right of fishing on reserves, 124; first set of regulations, 51, 54; and fish as common
property, 163; and fishing on reserves,
122, 123; on fishing privileges vs

Indian Act: forbidding of fund raising for title or land claims, 185; impact on Native peoples, 137; and Indian Affairs' control of lands and property, 124; and integration of Native people into immigrant society, 12; and Native reliance on cannery work, 143; and privileges on reserves, 123; and *R. v. Charlie,* 129; and reserve reductions and cutoffs, 181; status of Native peoples under, and lack of access to credit, 91, 138; and trespass on reserves, 122

Indian Affairs, Department of (Canada): 1923 meeting with Native leaders, 161-62; and access to fisheries, 195; Fisheries and, 60, 84, 99-100, 104, 105, 149-50, 180, 185-86, 195; and fishing on reserves, 2, 4, 123; and fishing rights, 27, 84, 101-2, 104-5, 195; and food fishing permits, 120-21; and foreshore fronting reserves, 152; jurisdiction over Indians, 33; on licensing of non-Native commercial fishers, 98; list of exclusive fisheries, 101; McKenna-McBride commission meets with, 171; Metlakatla Tsimshian appeal to, 72; and Native title, 104-5, 183; and Natives holding commercial licences, 137; Nisga'a petition to, 68; and non-Native seine licences, 149-50; and O'Reilly, 60, 61, 82, 83, 97; prohibition on raising of funds for Native title or land claims, 185; protection of Native fishing grounds, 82-83; and provincial government, 185; and *R. v. Charlie,* 2, 4; and reserves as self-sustaining, 39; and Sproat, 54; on transfer of jurisdiction over Native peoples in Confederation, 101-2

Indian Affairs Settlement Act, 181
Indian agents, 101, 111, 114, 125, 176, 177
Indian land question, 36-37, 54, 165, 166, 185, 187
Indian Reserve Commission, 43, 211n17
Indian reserve commissioners, 15; authority of, 85, 98-99, 180, 193-94; difficulty of work, 92; disappearance of position, 166; and Native testimony to

fisheries, 16. *See also* O'Reilly, Peter; Sproat, Gilbert Malcolm; Vowell, A.W.
Indian reserve commissions, 8, 16, 166, 187, 194. *See also* Indian Reserve Commission; Joint Indian Reserve Commission (JIRC)
Indian Rights Association, 166, 182
indigenous peoples: acquisition of land from, 11; imperialism and, 11; marginalization of, 11-12; settlement and, 9, 11
Institutes of the Law of England (Coke), 10
Interior Tribes Association, 166
Inverness cannery, 71
irrigation, 41, 42, 43, 45, 171
Issac, Joe, 119

Jack, R. v., 196-97, 225n29, 229n64
Jackson, Charlie, 174
Jacob, Allen, 193-94
Japanese fishers: alternative fleet supplied by, 142; as cannery employees, 163; cannery licences held by, 145, 156, 179; discouragement of, 15, 129; employment of, 154; Haida and, 161-62; numbers of, 135, 139; predominance of, 138
Jervis Inlet, 152-53
John, King, 81
Johnstone Strait, 157
Joint Indian Reserve Commission (JIRC), 36-42, 211n17. *See also* Anderson, A.C.; Indian Reserve Commission; McKinley, Archibald; Sproat, Gilbert Malcolm
Jonnie (Musqueam chief), 138
Joseph, Mathias, 138
Justice, Department of (Canada), 98-99, 152

Kamloops: Indian Reserve No. 2, 49-50, 51, 53(f); reserves, 30; trap fishing at, 42
Kanaka Bar, 49
Kelez, George B. *The Salmon and Salmon Fisheries of Swiftshore Bank,* 230-31n17
Kelly, Peter, 114-15, 121, 161-62, 182, 183-84, 196
Kennedy, Dorothy, 62(f), 63, 68

Watts, Dan, 172
Weaver, John, 19-20
weir fisheries, 38, 100; Cowichan River, 26, 41, 168, 169-70, 171; Powell River, 120
Wesley, John, 73
Wetmore, Edward L., 167-68, 170
Wet'suwet'en Indian reserve, 193
Whitcher, W.F., 31, 82, 83, 84, 85
White, Jacob, 155
White, Nathaniel W., 166-67, 173, 174
white fishers: in commercial fleet on Fraser River, 137; independent licences for, 140, 142-43, 177, 179, 190; numbers of, 135, 139; preferential treatment given to, 174, 176; as proportion of fleet, 135; seine- and trap-net licences, 155; seine licences for, 180, 190
Whyeek reserve, 49

Williams, J.C.T., 155, 171
Williams Lake, 63-64, 66
Wilmot, Samuel, 110, 131, 147, 148
Wilson, Richard, 76
Winche 7 reserve, 98
Windsor, C.S., 69-71
Winters v. United States, 239n26
Wit-at-Village, 100
women: as cannery employees, 163; employment in canneries, 142
Wright, Roland, 84-85

Yale, 56-57; fishery at, 30; salmon run and, 61
Yelakin, 46
Young, McGregor, 112, 169-70
Young, Robert J.C., 211n19

Ziff, Bruce, 79

Catherine E. Bell and Val Napoleon (eds.)
First Nations Cultural Heritage and Law: Case Studies, Voices, and Perspectives (2008)

Peggy J. Blair
Lament for a First Nation: The Williams Treaties in Southern Ontario (2008)

Lori G. Beaman
Defining Harm: Religious Freedom and the Limits of the Law (2007)

Stephen Tierney (ed.)
Multiculturalism and the Canadian Constitution (2007)

Julie Macfarlane
The New Lawyer: How Settlement Is Transforming the Practice of Law (2007)

Kimberley White
Negotiating Responsibility: Law, Murder, and States of Mind (2007)

Dawn Moore
Criminal Artefacts: Governing Drugs and Users (2007)

Hamar Foster, Heather Raven, and Jeremy Webber (eds.)
Let Right Be Done: Aboriginal Title, the Calder Case, and the Future of Indigenous Rights (2007)

Dorothy E. Chunn, Susan B. Boyd, and Hester Lessard (eds.)
Reaction and Resistance: Feminism, Law, and Social Change (2007)

Margot Young, Susan B. Boyd, Gwen Brodsky, and Shelagh Day (eds.)
Poverty: Rights, Social Citizenship, and Legal Activism (2007)

Rosanna L. Langer
Defining Rights and Wrongs: Bureaucracy, Human Rights, and Public Accountability (2007)

C.L. Ostberg and Matthew E. Wetstein
Attitudinal Decision Making in the Supreme Court of Canada (2007)

Chris Clarkson
Domestic Reforms: Political Visions and Family Regulation in British Columbia, 1862-1940 (2007)

Jean McKenzie Leiper
Bar Codes: Women in the Legal Profession (2006)

Gerald Baier
Courts and Federalism: Judicial Doctrine in the United States, Australia, and Canada (2006)

Avigail Eisenberg (ed.)
Diversity and Equality: The Changing Framework of Freedom in Canada (2006)

Randy K. Lippert
Sanctuary, Sovereignty, Sacrifice: Canadian Sanctuary Incidents, Power, and Law (2005)

James B. Kelly
Governing with the Charter: Legislative and Judicial Activism and Framers' Intent (2005)

Dianne Pothier and Richard Devlin (eds.)
Critical Disability Theory: Essays in Philosophy, Politics, Policy, and Law (2005)

Susan G. Drummond
Mapping Marriage Law in Spanish Gitano Communities (2005)

Louis A. Knafla and Jonathan Swainger (eds.)
Laws and Societies in the Canadian Prairie West, 1670-1940 (2005)

Ikechi Mgbeoji
Global Biopiracy: Patents, Plants, and Indigenous Knowledge (2005)

Florian Sauvageau, David Schneiderman, and David Taras,
with Ruth Klinkhammer and Pierre Trudel
The Last Word: Media Coverage of the Supreme Court of Canada (2005)

Gerald Kernerman
Multicultural Nationalism: Civilizing Difference, Constituting Community (2005)

Pamela A. Jordan
Defending Rights in Russia: Lawyers, the State, and Legal Reform in the Post-Soviet Era (2005)

Anna Pratt
Securing Borders: Detention and Deportation in Canada (2005)

Kirsten Johnson Kramar
Unwilling Mothers, Unwanted Babies: Infanticide in Canada (2005)

W.A. Bogart
Good Government? Good Citizens? Courts, Politics, and Markets in a Changing Canada (2005)

Catherine Dauvergne
Humanitarianism, Identity, and Nation: Migration Laws in Canada and Australia (2005)

Michael Lee Ross
First Nations Sacred Sites in Canada's Courts (2005)

Andrew Woolford
Between Justice and Certainty: Treaty Making in British Columbia (2005)

John McLaren, Andrew Buck, and Nancy Wright (eds.)
Despotic Dominion: Property Rights in British Settler Societies (2004)

Georges Campeau
From UI to EI: Waging War on the Welfare State (2004)

Alvin J. Esau
The Courts and the Colonies: The Litigation of Hutterite Church Disputes (2004)

Christopher N. Kendall
Gay Male Pornography: An Issue of Sex Discrimination (2004)

Roy B. Flemming
Tournament of Appeals: Granting Judicial Review in Canada (2004)

Constance Backhouse and Nancy L. Backhouse
The Heiress vs the Establishment: Mrs. Campbell's Campaign for Legal Justice (2004)

Christopher P. Manfredi
Feminist Activism in the Supreme Court: Legal Mobilization and the Women's Legal Education and Action Fund (2004)

Annalise Acorn
Compulsory Compassion: A Critique of Restorative Justice (2004)

Jonathan Swainger and Constance Backhouse (eds.)
People and Place: Historical Influences on Legal Culture (2003)

Jim Phillips and Rosemary Gartner
Murdering Holiness: The Trials of Franz Creffield and George Mitchell (2003)

David R. Boyd
Unnatural Law: Rethinking Canadian Environmental Law and Policy (2003)

Ikechi Mgbeoji
Collective Insecurity: The Liberian Crisis, Unilateralism, and Global Order (2003)

Rebecca Johnson
Taxing Choices: The Intersection of Class, Gender, Parenthood, and the Law (2002)

John McLaren, Robert Menzies, and Dorothy E. Chunn (eds.)
Regulating Lives: Historical Essays on the State, Society, the Individual, and the Law (2002)

Joan Brockman
Gender in the Legal Profession: Fitting or Breaking the Mould (2001)

Printed and bound in Canada by Friesens

Set in Galliard and News Gothic Condensed by Artegraphica

Copy editor: Audrey McClellan

Indexer: Noeline Bridge

Cartographer: Eric Leinberger